# Climate Change, Urbanization, and Water Resources

Heejun Chang • Alexander Reid Ross

# Climate Change, Urbanization, and Water Resources

## Towards Resilient Urban Water Resource Management

Springer

Heejun Chang
Department of Geography
Portland State University
Portland, OR, USA

Alexander Reid Ross
Department of Geography
Portland State University
Portland, OR, USA

ISBN 978-3-031-49629-5 ISBN 978-3-031-49631-8 (eBook)
https://doi.org/10.1007/978-3-031-49631-8



This Springer imprint is published by the registered company Springer Nature Switzerland AG
The registered company address is: Gewerbestrasse 11, 6330 Cham, Switzerland

Paper in this product is recyclable.

# Preface

This book project started during the pandemic in the summer of 2020 when the COVID-19 voluntary travel restriction was still in place in our town. Due to recommended social distancing, we met in Zoom many times to discuss this book project. The topic and the structure of the book were motivated by our life experience in some of our case cities and our teaching in Climate and Water Resources and Global Water Issues and Sustainability at Portland State University for many years. Drawing case studies from both global south and north, we seek to offer balanced perspectives while bringing both scientific knowledge in hydroclimatology and political ecology into water resource geography.

While climate change poses multiple threats to our society, the water resources sector is one of the most felt in our society and environment. With changing rainfall patterns and spatial and temporal distribution, water resource managers must rethink and reimagine future water resource management. In particular, since urban areas typically import water from surrounding rural areas, many cities are increasingly vulnerable to climate stresses. At the same time, as the urban population surpasses the rural population, and nearly two-thirds of the world population is projected to live in urban areas by 2050, it is imperative to understand water resource sustainability and resilient water resources management in cities across the globe.

We chose 12 case cities representing different stages of growth and economic development. As such, these cities illuminate different nexuses of histories and geographies that have shaped water resource infrastructure and management in each place. While global climate change impacts are felt globally, their local impacts vary depending on each city's social systems, ecological conditions, and technological infrastructure. Using the lens of the social-ecological-technological systems (SETS) framework, we show how humans manage water systems via soft and hard infrastructures and prepare for climate change adaptation. By comparing different case studies, we intend to draw some common lessons that others can consider when they revise their climate adaptation plan to make their water system resilient in the future.

During the book writing, we have witnessed unprecedented extreme weather events in our town. The September 2020 wildfire resulted in the evacuation of vulnerable populations in the Portland metropolitan region, while the June 2021 heat dome disproportionately affected the socially and economically underserved communities. The year 2023 is now considered the warmest year on record. We hope this book is useful for students majoring in geography, environmental studies, water resources, and urban planning as they seek balanced perspectives in both biophysical and social science aspects of water resource management. Additionally, this book will be helpful for practitioners whose interest intersects water resource management, climate change, and urban environmental planning. We appreciate the editorial team members at Springer whose comments are invaluable in helping clarify points in our manuscript. Any remaining errors solely belong to us.

Portland, OR, USA

Heejun Chang

Alexander Reid Ross

# Contents

# Chapter 1
# Introduction

**Abstract** This book discusses resilient urban water resources management in the context of climate change and ongoing urbanization. Twelve cities worldwide representing different climates and growth stages serve as case studies. Using these case cities, this book first identifies the main issues of water-related hazards in relation to the historical development of each city, investigates current strategies for dealing with climate-related water hazards, and explores potential adaptive strategies. By comparing and contrasting these case studies, we plan to draw some common lessons while acknowledging place-based distinctive strategies.

**Keywords** Climate change · Urbanization · Floods · Droughts · Resilient water management

As climate change-related extreme events become more common around the world, so do the impacts on our society and environment. The past four decades have witnessed such extreme events, inundating major human population centers with unprecedented intense precipitation and limiting water consumption resulting from prolonged droughts. In the United States, for example, there have been 341 weather and climate-related disasters since 1980, with estimated overall damages/costs exceeding $2.475 trillion (NOAA 2023). Globally, the population exposed to floods has increased substantially by 58–86 million from 2000 to 2018 (Tellman et al. 2021).

Floods and droughts are two of the most damaging natural disasters, resulting in socioeconomic, ecological, and infrastructure stresses. While substantial regional variations exist, according to the latest climate change projections, continuous warming throughout the twenty-first century is highly likely to lead to earlier snowmelt and subsequent spring floods in higher mid-latitudes. Rainfall intensity in some regions has also increased, further exacerbating urban flooding. At the same time, flash droughts could become more common as soil moisture could dry up quickly in a warming atmosphere (Caretta et al. 2022).

Of particular concern are urban regions where high amounts of impervious surface areas do not absorb excess runoff and increase downstream flooding while lowering baseflow during the dry period. With the conversion of natural areas into



H. Chang, A. R. Ross, *Climate Change, Urbanization, and Water Resources*,
https://doi.org/10.1007/978-3-031-49631-8_1

concrete surfaces, many world cities face urban heat islands. As a result, urban regions face the dual challenges of additional warming and changing hydrology. Additionally, many of the world's major cities are located in coastal areas. With sea level rise and storm surges, these coastal cities are prone to flooding. As the urban population is continuously expected to grow throughout the twenty-first century (United Nations 2022), it is imperative to assess how urban regions could be more vulnerable to climate-induced hazards.

From 1950 to 2020, the proportion of the global population living in cities jumped from 25 to 50%. While urban regions are major sources of greenhouse gas emissions and are vulnerable to climate-induced water hazards, they are also places of climate mitigation and adaptation through innovative interventions since many such decisions are made at the local level. Since many urban practitioners are keenly interested in developing various climate mitigation and adaptation strategies, comparing and contrasting multiple strategies across different cities help city land and water planners evaluate their strategies to tackle climate challenges.

Water resource impacts of climate change vary substantially by region because of not only the projected changes in precipitation and temperature but also the way water infrastructure is developed in each city (Chang 2019). Development seldom occurs on a linear timeline. There are quantum leaps and massive setbacks; destruction of flood controls by natural disasters causes devastation to entire cities; revolutions turn customs and traditions on their head, making and remaking existing practices and structures; and the fall of empires leaves behind piecemeal systems meant to serve colonial interests rather than public ones.

The uneven distribution of water delivery and flood infrastructure can expose different groups to different vulnerabilities within the city. There is a growing consensus among hydrologists showing that water resources and human populations have coevolved over time (Sivapalan et al. 2012), such that imagining the hydrologic cycle outside of human influence, and vice versa, becomes abstract and immaterial. Therefore, assessing the current relationship between water and the city requires contextualization within a hydrosocial and/or socio-hydrological framework, with the former indicating more a theoretical corpus on power relations and the latter a holistic approach to engineering water infrastructure that accounts for and includes the work of human societies (Ross and Chang 2020).

The postcolonial condition began with newly forged social relations amid discordant, fragmentary, and underdeveloped infrastructure, complicating the coherence of modern nation states. In much of the Global South, the complexity of jarring temporal and spatial change, mixed with the challenges of facilitating societal solidarity and the hierarchical tradition of civil engineering, brought about the problem of hydrosocial coordination outside of bureaucratization and technocratic management (Molle et al. 2009; Niranjana 2021).

At the same time, the urgency of the problem of climate change calls for broad socio-technological transformation inclusive of decentralized strategies that, in many cases, appear effective in managing the impacts of floods and droughts. We use 12 case studies around the world to draw how each city relates to climate change and water resource management. These cities, comprising seven coastal and five inland cities, represent various physical and social settings and various stages of urban development (Fig. 1.1).

These cities do not show “leaders and followers,” but the complexity of different approaches to water resource management in the urban context, ranging from “water-sensitive cities” to the Istanbul consensus, inviting such approaches as sustainable urban drainage systems and runoff attenuation features, while also incorporating larger infrastructure projects involving buffer islands, canalization, and dams. While cities of the Global South often utilize larger infrastructure projects, in part resulting from the postcolonial legacy, increasing efforts involving a shift toward participatory planning in cities such as Lagos suggest renewed policies of inclusion. At the same time, problems remain in cities of the developed world, which despite successful decentralization of flood control sometimes use the appearance of inclusion in order to promote techno-fixes such as desalination without sufficient popular support.

Our case studies reveal one undeniable truth: All countries are threatened by climate change, and poorer nations often face the greatest challenges, both in terms of impacts and mitigation efforts. The fiercer the challenges, the less likely communities can remain stable and creative over time. The development of water infrastructure is ultimately not only about outcomes but also about inputs; in other words, it means creating an equitable product to build civic participation. Systems include feedback mechanisms, and delivery or nondelivery of water creates a major impact on the building of communities that can be resilient to climate hazards or brittle and resistant to change.

Due to the troubling correlation between climate impacts and poverty, coupled with the global need to mitigate climate change, the international community (or communities) should engage more in developing responsive water infrastructure,

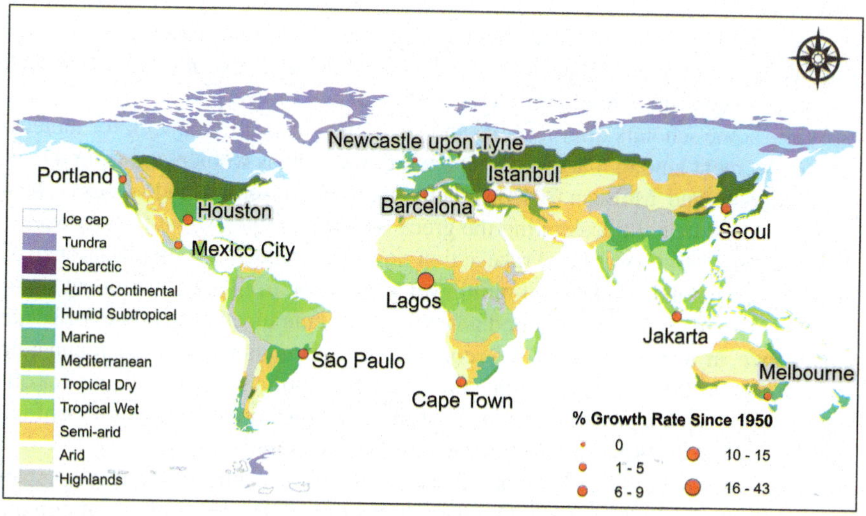


**Fig. 1.1** Location of study sites with climates and population growth rates

such that positive feedback can promote stability and growth instead of perpetual crisis. This means developing more holistic ways of understanding atmospheric phenomena as deeply connected to the other spheres of the earth system, including land use, for instance, in conceptual and technical models of climate change. By integrating human agency with that of the physical environment, greater understanding can be attained toward modifications of social behaviors that would benefit human populations and curb climate change at the same turn.

By studying how water systems develop in tandem with water-related hazards, we found not only that human–water systems are generally coevolved but that their specific coevolution in time and space springs from the metabolic processes of those places and further produces societal interactions, ways of belonging, power relations, epistemologies, and understandings of territory. Due to these specificities of experience, adaptation strategies emerge differently across different study areas, particularly as it pertains to the structuring of decisions, hierarchies of decision-making power, and interest in participatory adaptive management by decision-makers or by the general public.

Here, we assert that the "long-range" view helps assess present potentialities, not due to path dependence or socio-economic tracks but in relation to the dialectics of continuity and rupture, an often hybridized and multilayered process of abrupt change and consistent processes. In this way, we can take into account the legacy effect of influences and effects of past policies and historic movements, while also predicting more beneficial and productive strategies suited to the appropriate context (Table 1.1).

As noted above, we found different adaptation strategies for different countries, with wealthier countries often taking approaches closer to decentralized participatory management, including the water-sensitive cities platform, as well as investments in broader instrumentation for quicker responses to and closer knowledge of water-based climate hazards. Poorer countries, on the other hand, appear to experience sharper consequences of climate hazards and adopt mitigation strategies that can take a more grand and sweeping form, while their populations often lack resources to contribute to decentralized and voluntary projects.

While some scholars offer a social and ecological justice framework for increasing involvement among marginalized populations, such as women and slum dwellers (Adjibade and McBean 2014), others, such as consulting and advisory firm Deloitte, insist that "[a]ccelerating the green transition requires that C-suite executives lead the path forward and take action now" (Deloitte 2023). In places such as Newcastle, England, alliances of multilevel authorities, from the government to the private sector to civil society, found a way to contribute together to a kind of social renaissance in a city depleted by economic transformations, helping recuperate neglected areas for experimentation and development of adaptive strategies for climate change. Such a program can be generalized throughout the major cities of the world, producing a brilliant future for generations to come.

There is no silver bullet, but every place has a different and specific way to adapt to local conditions in keeping with a global effort to confront the effects of climate change. Of course, the other side of this is dealing with the causes. Climate effects

**Table 1.1** Study cities' climate, climate-related hazards, and urban growth characteristics

| City (C = coastal, I = Inland) | Climate | Projected flood risk (IPCC 2022) | Projected drought risk (IPCC 2022) | Growth stage | Metropolitan population (millions) | Water-related hazards (* mild, ** severe, *** extreme) (water supply, floods) |
|---|---|---|---|---|---|---|
| Seoul (I) | Humid continental | High | Moderate | Declining | Mega | ** |
| Jakarta (C) | Tropical | High | Very high | Growing | Mega | *** |
| Melbourne (C) | Temperate oceanic | Low | Moderate | Stabilized | Large | *** |
| Istanbul (C) | Temperate Mediterranean | Low | High | Stabilized | Mega | ** |
| Barcelona (C) | Mediterranean | Low | High | Declining | Large | * |
| New Castle (I) | Marine west coast | Low | High | Stabilized | Regional | ** |
| Lagos (C) | Tropical | High | High | Growing | Mega | *** |
| Cape Town (C) | Mediterranean | Low | High | Stabilized | Large | *** |
| São Paulo (C) | Humid subtropical | High | High | Growing | Mega | ** |
| Mexico City (I) | Subtropical highland | High | High | Growing | Mega | ** |
| Houston (I) | Humid subtropical | Low | Moderate | Growing | Large | *** |
| Portland (I) | Mediterranean | Low | Moderate | Stabilized | Mid-sized | * |

will worsen unless carbon emissions are brought down, and global trends do not indicate a global decrease in emissions at anything near the scale necessary to contend with the developing problem. Many cities in our study will see impacts from rising sea levels caused by the thermal expansion of oceans and the melting of glaciers, as well as changes in the water cycle. Without lowering emissions, more glaciers will melt, reducing the albedo, or reflectivity, of the Earth's surface and thus committing it to further warming. Feedback like this, methane release through permafrost melting, and others will bring about changes beyond present adaptive capacity, while the chemical buildup of carbon in the atmosphere will also have secondary effects, such as ocean acidification, for which no technological fixes exist.

For all of these reasons, safe planning for climate change with regard to water resources in cities requires deep consideration and analysis of both impacts and causes. In Houston, for instance, large-scale efforts are underway to reclaim wetlands and forests that could serve as breaks against tidal storm surges and hurricanes. From land use changes to sustainable infrastructure and design, urban water issues can be approached on a multi-scalar basis—starting with small actions such as rain gardens and bioswales to deter pluvial flooding, and planting fruit trees and vegetables for community gardens. Enhancing community engagement, creating afterschool activities, involving youth from kindergarten through high school, and thus developing leadership positions to inspire people and educate their parents—these things can cultivate social change on larger scales and build democratic systems to increase participation in civic life.

To develop these projects and increase participation, people need access to resources such as community programs that can help them engage in their own time; they need incentives to change the structure of the city from house to house, forming decentralized adaptive systems themselves. By setting up civic infrastructure to attract and recruit local communities to increase political participation through outreach, education, and resource exchanges, cities can empower people to change unsustainable structures of everyday life and create new resources for themselves. In this way, social, ecological, and technological systems (SETS) all work in tandem (Marklof et al. 2018), as part of an integrated system in which technological exchanges and social systems match technological developments to improve human relationships with the lived environment, increasing the quality of life, supporting flexible and adaptive communities, and mitigating both the causes and effects of climate change.

Yet, stopping at the government level cannot fulfill the needs of society. Relying on the public to actively engage in autonomous self-management without any support remains similarly idealistic. While each actor contributes to resilience in their own ways, through a strong integration of academics and policymakers, the coproduction of knowledge can result in actionable changes and increase the effectiveness of resilience. Academic programs can promote safe and well-documented experimentation ethically integrated with community growth and developed in a generalizable way that can benefit society. We need to create models for adaptive improvement that avoid the pitfalls of gentrification—such as lowering interest rates for loans specifically targeting climate adaptation and deepening the sense of place in neighborhoods.

Flood mitigation requires collaboration across spatial connectivity (uphill to downhill) and connectivity across stakeholders. Through our studies, we found that adaptation often follows disaster, and to break that cycle, it is important to develop intellectual and social networks that can better coproduce knowledge in predicting hazards before disasters arrive. As the example of Barcelona shows, adaptive strategies can become neighborhood events that bring neighbors together to comingle—community events that draw together larger parts of the city. The Istanbul Consensus shows, on a larger scale, that scientific consensus can build toward policy initiatives, and forums, conferences, and public venues can hold organizing potential for enlightened discussion.

More generally, we found that the "three C" model remains imperative: Communication leads to collaboration, which is necessary to make change happen. Where urban–rural divisions lead to less communication, collaborative and participatory methods become more difficult.

Transparent communication can further improve community integration, as shown by the Melbourne example, but without it, good ideas can be compromised. Technology imposed from above rarely empowers people. The interweaving of traditional indigenous knowledge and community-based technologies can help build stability among vulnerable populations and ensure that resilience spreads from the ground up, as the case study of Lagos shows.

All of this goes to show that, while top-down models tend to be less functional than bottom-up models, an overreliance on participation without adequate civic infrastructure renders the latter unstable. In contrast, models that target small projects with sufficient participation may lack the scale necessary to adapt comprehensively to climate change impacts. Thus, in all our study areas, we find that integration, conceptual vision, and wide-reaching reform go hand in hand, offering the best possibilities not just for resilience to climate hazards but for transformative practice that can address both causes and effects. Here, the reclamation of landscape, the easing of zoning restrictions to make way for green infrastructure, the development of new urban models, and the allaying of individual action and community development demand an ethos of cooperation in order to adapt to the future of climate change.

## References

Ajibade I, McBean G (2014) Climate extremes and housing rights: a political ecology of impacts, early warning and adaptation constraints in Lagos slum communities. Geoforum 55:76–86

Caretta MA, Mukherji A, Arfanuzzaman M, Betts RA, Gelfan A, Hirabayashi Y, Lissner TK, Liu J, Lopez Gunn E, Morgan R, Mwanga S, Supratid S (2022) Water. In: Pörtner H-O, Roberts DC, Tignor M, Poloczanska ES, Mintenbeck K, Alegría A, Craig M, Langsdorf S, Löschke S, Möller V, Okem A, Rama B (eds) Climate change 2022: impacts, adaptation and vulnerability. Contribution of working group II to the sixth assessment report of the intergovernmental panel on climate change. Cambridge University Press, Cambridge/New York, pp 551–712. https://doi.org/10.1017/9781009325844.006

Chang H (2019) Water and climate change. In: Richardson D, Castree N, Goodchild MF, Kobayashi A, Liu W, Marston RA (eds) International Encyclopedia of geography people the earth environment and technology. Wiley, pp 1–6

Deloitte (2023) Deloitte 2023 CxO sustainability report. https://www2.deloitte.com/content/dam/Deloitte/ch/Documents/risk/2023-deloitte-cxo-sustainability-report.pdf

IPCC (2022) In: Pörtner H-O, Roberts DC, Tignor M, Poloczanska ES, Mintenbeck K, Alegría A, Craig M, Langsdorf S, Löschke S, Möller V, Okem A, Rama B (eds) Climate change 2022: impacts, adaptation, and vulnerability. Contribution of working group II to the sixth assessment report of the intergovernmental panel on climate change. Cambridge University Press, Cambridge/New York, p 3056. https://doi.org/10.1017/9781009325844

Markolf S, Chester M, Eisenberg D, Iwaniec D, Ruddell B, Davidson C, Zimmerman R, Miller T, Chang H (2018) Interdependent infrastructure as linked social, ecological, and technological systems (SETS) to address lock-in and improve resilience. Earth's Future 6(12):1638–1659

Molle F, Mollinga PP, Wester P (2009) Hydraulic bureaucracies and the hydraulic mission: flows of water, flows of power. Water Altern 2(3):328–349

Niranjana R (2021) An experiment with the minor geographies of major cities: infrastructural relations among the fragments. Urban Stud 59(8):1556–1574

NOAA National Centers for Environmental Information (NCEI) (2023) U.S. Billion-Dollar Weather and Climate Disasters. https://www.ncei.noaa.gov/access/billions/. https://doi.org/10.25921/stkw-7w73

Ross A, Chang H (2020) Socio-hydrology with hydrosocial theory: two sides of the same coin? Hydrol Sci J 65(9):1443–1457. https://doi.org/10.1080/02626667.2020.1761023

Sivapalan M, Savenije HHG, Blöschl G (2012) Socio-hydrology: a new science of people and water. Hydrol Process 26(8):1270–1276. https://doi.org/10.1002/hyp.8426

Tellman B, Sullivan JA, Kuhn C et al (2021) Satellite imaging reveals increased proportion of population exposed to floods. Nature 596:80–86. https://doi.org/10.1038/s41586-021-03695-w

United Nations, Department of Economic and Social Affairs, Population Division (2022) World Population Prospects 2022: Summary of Results. UN DESA/POP/2022/TR/NO. 3

# Chapter 2
# Seoul, Republic of Korea

**Abstract** Developed as the capital of the Yi Dynasty in the late fourteenth century, Seoul's long-term rainfall measurements show substantial high interannual precipitation variability with significant increases in summer precipitation intensity in the last 30 years. As one of the fastest growing cities worldwide in the latter half of the twentieth century, intensive land development, mostly converted from rice paddies or wetlands, has resulted in chronic flooding in low-lying areas along the Han River and its tributaries. To reduce flood risks, the city government has heavily relied on technological solutions such as levees, water pumping stations, and underground storage. Thanks to these structural measures, riverine flooding has been somewhat reduced in recent decades. However, localized intense rainfall with a changing climate has resulted in urban flooding in many developed city areas. One promising aspect of Seoul's flood risk management is that, with active citizen participation thanks to a high education rate, the city now appears to have embraced more social and ecological dimensions by restoring streams, installing green infrastructure, and social education, which will become increasingly more important for comprehensive, equitable, and just flood risk management.

**Keywords** Flood risk management · Development · Urban flooding · Stream restoration · Seoul

## 2.1 Introduction

In late July 2011, a record-breaking approximately 500 mm of heavy rainfall lasted 2 days in Seoul, South Korea. In downtown Seoul, the storm created flash floods, inundating the main street in front of the city hall and halting the heavy traffic during rush hours. In the southern part of the city, the heavy storm triggered landslides, collapsing apartment building structures. Landmines from the Korean Civil War in adjacent mountains had been buried with debris flows, creating further anxiety among residents near the landslide site. Together, this rain event resulted in the loss of 69 people, with more than 125,000 people affected. In response to this extreme


H. Chang, A. R. Ross, *Climate Change, Urbanization, and Water Resources*, https://doi.org/10.1007/978-3-031-49631-8_2

event, the city revised its flood risk management plan while establishing a dedicated flood research center in a city-sponsored research university. As a result, the design standard for storm pipes has changed, while underground water tunnels were constructed to capture excessive stormwater during flood events. Despite these efforts, these new pipes and underground tunnels did not prevent many parts of the city from being inundated with intense rainfall in August 2022 (141 mm per hour with 434 mm total precipitation in 24 h).

As an old city that expanded rapidly after the Korean Civil War, Seoul's intensive land development in low-lying areas further exacerbated riverine and urban flooding problems, by removing wetlands and floodplains. Since Seoul became one of the most environmentally conscious cities in the world, the Seoul government has strived to change its core environmental strategies from purely technology-based solutions to ecologically sensitive and socially responsive strategies. In response to growing citizen interests, new environmental plans have been coproduced with diverse stakeholders' input. Taking advantage of highly skilled and educated laborers, the city has sought to harmonize nature with modern technology to create humane solutions to tackle climate-related problems. For example, like most developed cities in Western societies, major stream restoration projects have been largely successful with the active participation of citizens. With respect to the citizen-led climate action plan, as part of the Cities100 project, Seoul had an ambitious goal of reducing greenhouse gas emissions by 25% from 2015 to 2020. However, the rapidly changing climates with a more frequent intensity of rainfall and heat challenged these endeavors to cope with water-related hazards.

## 2.2 Climate and Geography

Located in the convergence zone between the Asian continent and the Pacific Ocean in mid-latitude, Seoul has four distinct seasons. Siberian high pressures bring strong wind and cold air to the Korean peninsula in winter, while warm maritime air from the Pacific Ocean brings wet and humid summers. The annual precipitation is highly concentrated in the summer monsoon season (June–September). Approximately 70–80% of annual precipitation falls during this monsoon season. Together with monsoonal rainfall and typhoon-induced precipitation, floods are common during the summer months.

Invented during King Sejong in the Yi dynasty, *Cheug-ugi* in the mid-fifteenth century, a rainfall measuring device, provided a long-term rainfall record in Seoul for nearly five centuries. Like most places in mid-latitudes, there exists a substantial interannual variability of annual precipitation, with some years of precipitation amounting to more than 2000 mm while other years have as low as 300 mm. However, summer precipitation extremes and intensity have increased in the last century. Using the *Cheug-gi* and modern rain gauge data, Choi (2016) showed that total precipitation amount and extreme precipitation events increased in summer in Seoul, particularly since the 1990s. The summer monsoon season's rainfall distribution has become less variable, with the disappearance of the later break period (late July to early August) due to increased precipitation in the later break period. Such changes were associated with the expansion of subtropical high-pressure cells due to rising ocean temperatures in the Western Pacific and the advection of high-pressure air masses toward East Asia. The confluence of these two different air masses enhanced rising motions, which then were likely to be linked to the increases in the mean and extreme precipitation events (Choi 2016).

Mean annual precipitation increased from 1021 mm to 1646 mm between 1910 and 2012. During the same period, air temperature increased by 1.9 °C, indicating that such an increase is attributed to global climate change in Seoul (Seoul Research Institute 2013). In more recent years, from 1973 to 2022, maximum hourly precipitation increased substantially, even though summer mean precipitation slightly increased. This increasing maximum hourly precipitation is attributed to the increase in intensity and frequency of extreme precipitation (Kim et al. 2023). While one could argue that urbanization contributed to increases in air temperature, the relative contribution of urban development to increases in air temperature is less than 50% in major cities in South Korea, suggesting that global climate change is more responsible for increasing air temperature (Fig. 2.1).

As the capital of the *Yi* dynasty for over 500 years, Seoul was founded in a relatively safe place protected from strong winds and overflow of water from streams. According to Korean geomancy, *Pungsui* (wind and water), downtown Seoul, surrounded by Bukhan mountain in the north and Cheonggye stream in the south, was ideal for shielding the northwesterly wind from the north in winter and preventing frequent flooding in summer. It was the very reason that the first king Yi Seong-gye,

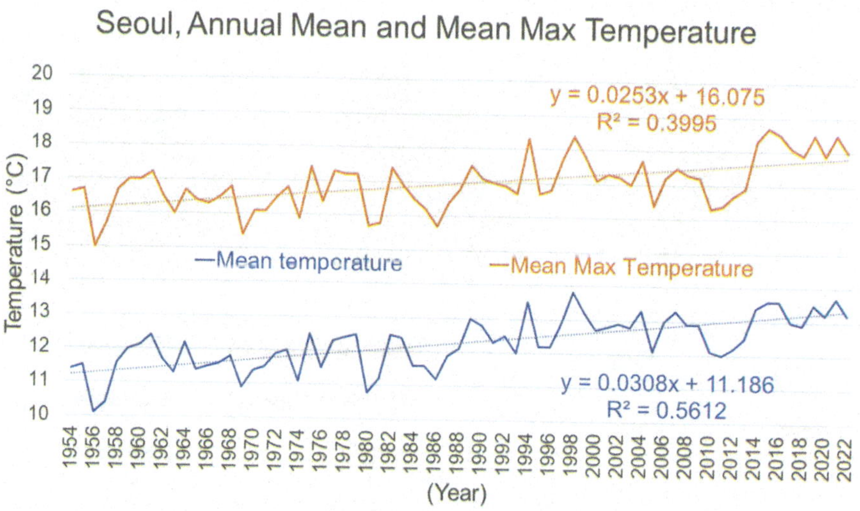


**Fig. 2.1** Changes in annual mean and annual mean maximum temperatures in Seoul, 1954–2022. (Data source: Korea Meteorological Administration, 2023)

with the help of jikwan, who evaluated the good fortune of a place based on *Pungshu*, selected the place for the capital of the Yi dynasty.

As summer rainfall brought excessive sediments from surrounding regions, the kings regularly dredged the bottom of the Cheonggye stream (meaning Clean stream), so it did not overflow into surrounding areas. Those kings who managed mountains and rivers well were considered to be good kings in ancient days. As Korea became a colony of Japan in 1910, the city grew rapidly, with many migrants settling along the Cheoonggye stream. Once a place for swimming, fishing, and recreation, the stream became degraded with wastes from nearby houses where poorest people resided.

In the modern era, mainly in the twentieth century, the city expanded multiple times to absorb surrounding agricultural lands and forests to accommodate the growing population. Lands once used for rice paddies were converted into residential, industrial, or commercial areas. The most substantial land development has occurred since the end of the Korean Civil War in the early 1950s. Migrants from north and southern provinces not only increased the urban population but also created typical urban ghettos near downtown areas. In the vicinity of downtown Seoul, the hilly regions were rapidly developed for settlements for rural migrants, with many of their houses built without building permits. These areas also did not have drinking water supply or sewer systems, creating downstream pollution. Since major corporations' headquarters and governmental sectors were concentrated in Seoul, its land development further accelerated until the late 1980s. According to land cover analysis between 1975 and 2010, urbanized areas increased from 165.6 $km^2$ (27.4%) to 341.7 $km^2$ (57.8%), while agricultural land areas decreased from 202.5 $km^2$ (33.5%) to 22.2 $km^2$ (3.7%) (See Fig. 2.2).

As a result of major new town developments in the suburban region of Seoul, Seoul's population started to decline in the early 1990s, losing nearly a million

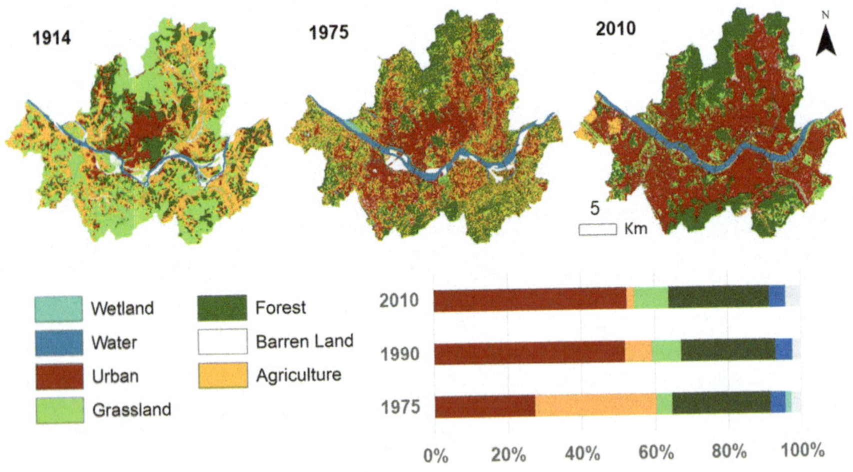


**Fig. 2.2** Changes in Seoul's land cover from 1910 to 2010. The bar graph shows changes in each land cover between 1975 and 2010

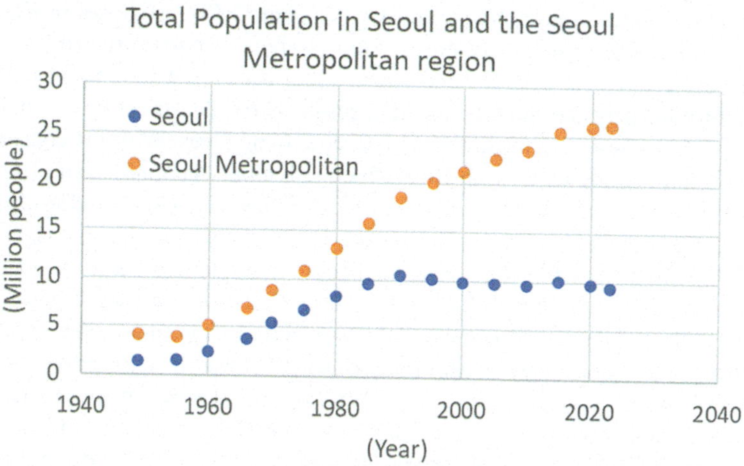


**Fig. 2.3** Population growth in Seoul and the Seoul Metropolitan region, 1949–2023. The Seoul metropolitan region includes Seoul, Incheon, and the Gyonggi province. (Data source: Korean Statistical Information Service (kosis.kr))

people by 2020. In contrast, the population of the Seoul metropolitan region has increased sharply since the 1990s (Fig. 2.3). These contrasting population trends are attributed to the migration of Seoul residents to new suburban cities that have been developed since the 1990s. With the aging population and the lowest fertility rate in the world, the country's population is continuously projected to decline in the future, while it is unclear if Seoul's population will remain the same.

This conversion of natural areas to built environments resulted in substantial alterations in urban hydrology. As impervious surface areas increase, peak runoff time is reduced, increasing more frequent urban flooding. Without coordinated land use planning, the haphazard development has further exacerbated flooding programs, particularly in low-income neighborhoods located in the vicinity of urban streams and the Han River (Bae and Chang 2019). At the same time, together with the mining of groundwater, decreases in the infiltration of rainwater to the soils and declines in baseflow and groundwater levels, threatened the region's water security (Lee et al. 2018). In Seoul, many low-lying areas are prone to flooding with a precipitation of 50 mm/h.

## 2.3 Water and Land Management in Seoul

Since the 1950s, Seoul's water resource management has gone through different cycles. In the initial era (1950–1980s), the development paradigm dominated. In aspiration for economic prosperity, the period coincides with the massive migration of rural people to Seoul. Many local streams disappeared either by burial or rerouting during the development era (Chang et al. 2020, as shown in Fig. 2.4). Sewers were not separated from streams; thus, to remove odors from streams, most streams were covered. The covered streams had been used for roads or parking spaces in a densely developed city. As a result of the densification of land uses, flooding was common in most downstream low-lying areas. With the construction of major dams in the upper Han River, flood risk along the main stem of the Han River was substantially reduced. In contrast, flooding in small tributaries, particularly in newly developed areas, increased. For example, as the movie *Parasite* shows, old slums in low-lying areas with people living in half sub-basement (*banjiha*) had constant flooding problems. The construction of *banjiha* was encouraged in the 1970s and 1980s during the Cold War period to minimize the damage of potential military attacks by North Korea. However, it was the poor who became the victims of urban floods as those living in *banjiha* got trapped in water when their homes got flooded with water since *banjiha* essentially served as a collection point of flood water. The legacy persists today. In August 2022, when there was torrential rainfall (over 300 mm/day), many houses in low-lying regions got inundated.

Another notable housing development in Seoul in this period was the initiation of a mass supply of high-rise apartment buildings. Realizing that providing single-family detached houses alone was not sufficient to supply enough housing to the rapidly growing population, the government promoted mass housing development by constructing high-rise apartment buildings, typically tightly connected to big private companies. These new mass apartment buildings were concentrated in southern districts, such as Gangnam district, where most vacant or agricultural lands were present during the period. As shown in Fig. 2.4, meandering stream channels have been severely modified with stream straightening resulting from the conversion of agricultural lands to residential or commercial development.

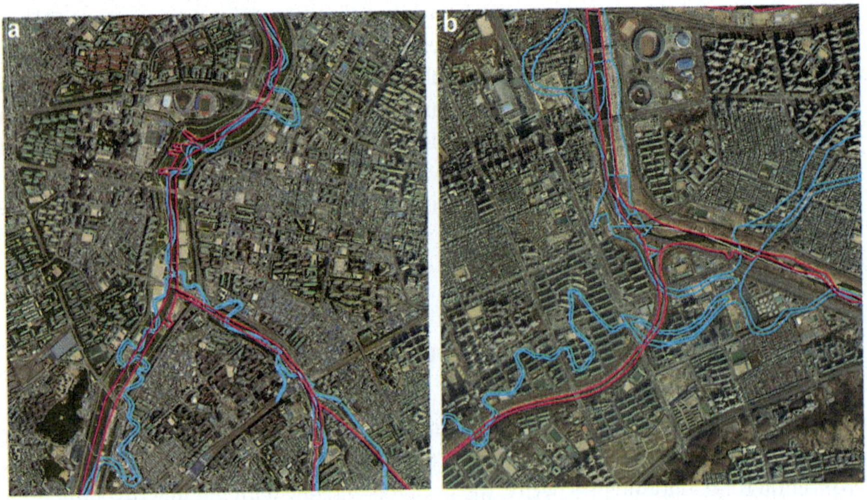


**Fig. 2.4** Historical (blue) and current (purple) streamlines in (**a**) Anyang-chun in Southwestern Seoul and (**b**) Tan-chun in Southeastern Seoul

The second phase is characterized as a transition period, from the 1990s to the early 2000s. This period coincides with the transfer of power from the authoritarian military government to the democratic government in South Korea. Additionally, the central government transferred some core governmental functions to the local government in the mid-1990s, allowing the local government to initiate and implement new projects. Together with the newly established Ministry of Environment, local grassroots organizations demanded new approaches to land and water management, which planted seeds for major stream restoration in the subsequent years. While land development in Seoul was somewhat stabilized, with new town development in the vicinity of Seoul, many agricultural lands were converted to intensive apartment housing. Such development resulted in water pollution and downstream flooding of tributaries flowing into Seoul (Chang 2008).

The third phase is the technology-based stream restoration era. When Lee Myung-bak became the new mayor of Seoul in 2003, he had a bold plan to open the covered stream, namely the Cheonggye stream (Fig. 2.5). While the stream was covered to create an intracity highway and promote inner-city development, the surrounding areas soon became degraded with limited views and noise from the highway. The two-floor highway, built on stream, was also too old to maintain its function without major repairs. To revitalize the region and fix the aging infrastructure, the mayor then initiated a major stream restoration project by removing the highway. While debated due to its main focus on the economic revitalization of downtown by removing aging infrastructure during the project period, once completed, the Cheonggye stream restoration project provided some ecosystem benefits to the surrounding area such as temperature reduction and recreation for urban residents in the middle of skyscrapers. Following this major downstream stream restoration project, many other districts in Seoul initiated uncovering streams, creating places for aesthetics and recreation.

While the opened streams provide trails for biking and walking, they also become vulnerable places during rainfall events. Considering that most rainfall events are localized and streams initiate from mountainous areas with steep slopes, these urban streams are very flashy, increasing water levels in a short period of time. Without adequate warning and evacuation routes, several people died in some tributaries during summer storm events.

## 2.4 History of Flooding and Flood Risk Management in Seoul

Seoul's biggest flood in the twentieth century occurred in 1925 when two consecutive typhoons passed through the region 4 days apart. With cumulative precipitation amount nearing 650 mm, many levees along the Han River broke, inundating not only the adjacent areas of the river but also as far as downtown Seoul. Approximately, 400 people died, while more than 12,000 houses were destroyed. After this big

**Fig. 2.5** Restored Cheonggye stream in downtown Seoul and a flood warning sign for closing trails during rain events. (Source: Photo by author)

flood, the Japanese Governor-General of Korea designed a revised water master plan for the Han River, constructing levees and installing water pumping stations along the mainstem Han River from 1926 to 1935. Additionally, the Japanese Governor-General of Korea also constructed levees along the other major tributaries of the Han River. In the mid-1960s, another major flood hit Seoul for two consecutive years, prompting a major river water management project. While constructing levees along the Han River, highways were constructed on these constructed levees. The city then reclaimed lands adjacent to the highways, building major apartment complexes. The construction of these roadways and buildings reduced floodplains along the river and changed the river's flow paths, resulting in constant flooding in these reclaimed lands (Jung 1990) (Table 2.1).

Urban fluvial flooding was a consistent problem in the 1980s and 1990s. The city experienced two major floods in subsequent years in the mid-1980s and early 1990s. In the 1980s, the major floods of Seoul in 1984, which affected more than a quarter million people, the Seoul government initiated a major river maintenance project along the Han River (see Fig. 2.6). This effort coincided with a preparation plan for the Asian Games in 1986. To improve river flow and remove fragmented wetlands along the floodplain in the Han River, the city constructed a sloped brick all along the mainstem Han River. As a result of the river maintenance project, floodplain areas were converted into recreational spaces containing trails, outdoor swimming pools, and parking lots. Unfortunately, such a maintenance project has reduced the

**Table 2.1** Major flood years in Seoul, precipitation amount, and damage, 1925–2022

| Year | Precipitation | Period | Dead & missing people | Refugees |
|---|---|---|---|---|
| 1925 | 500 mm | 7/16–7/18 | 647 | Unknown |
| 1984 | 334.4 mm | 9/3–9/5 | 43 | 289,800 |
| 1990 | 486.2 mm | 9/9–9/11 | 44 | 93,200 |
| 1998 | 332.8 mm | 8/8 | 24 | 2290 |
| 2001 | 273.4 mm | 7/15 | 139 | 470 |
| 2010 | 302.5 mm | 9/19–9/21 | 3 | 41,600 |
| 2011 | 587.5 mm | 7/26–7/28 | 22 | 34,300 |
| 2012 | 75.5 mm | 7/13 | 0 | 767 |
| 2013 | 279.5 mm | 7/11–7/15 | 1 | 160 |
| 2018 | 497 mm | 8/28–8/30 | 0 | 117 |
| 2022 | 515.5 mm | 8/8–8/9 | 8 | 205 |

Source: Seoul Metropolitan government

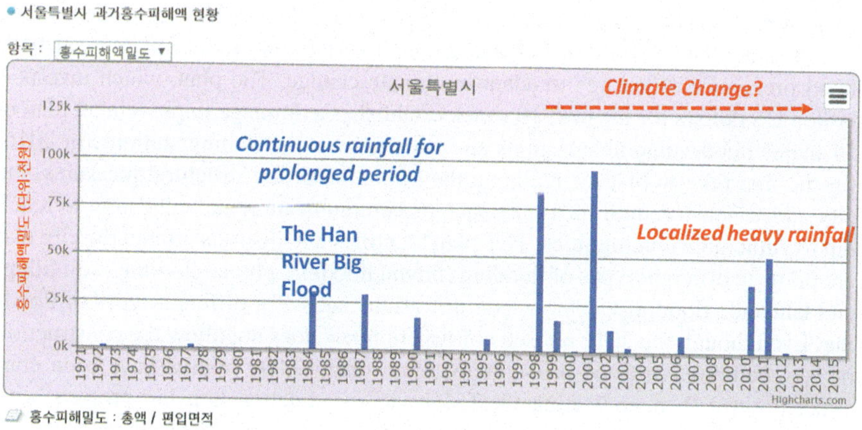


**Fig. 2.6** Flood damage and major flood events in Seoul. (Source: Korea Ministry of Interior and Safety 2022)

capacity of the river to hold water during high flood events. During the 1980s, additional focus was given to installing sewer pipes that could separate sewers from main streams to improve degraded water quality. As shown in these examples, a technological approach dominated the paradigm of flood risk management in Seoul.

In the summer of 1990, the second biggest flood in the twentieth century happened in the lower section of the Han River. Caused by a typhoon, the flood resulted in 44 dead or missing people, and over 90,000 people displaced. Several levees broke, inundating many residential and agricultural lands adjacent to the Han River. One of them was located in the lower Han River, which was constructed after the 1925 flood to prevent the overflow of river water to adjacent agricultural lands. The water level in several neighborhoods of the Gangdong district was as high as 1.5 m, inundating the district's office building. About 800 people were isolated by the rising water. This place was the old floodplains with wider channels, absorbing excessive water during flood events. With limited water-holding capacity, the newly developed areas became the victims of floods. After this flood, the Seoul city government established a 3-year flood hazard plan by installing 28 new rainwater pumping stations, strengthening levees, and deepening channels for 30 km.

Despite these efforts, the city was again devastated by big floods in 1998 and 2001. These two floods collectively resulted in more than 160 dead or missing people. The Seoul city government extended the flood hazard plan for another 5 years, installing 19 new rainwater pumping stations, strengthening levees for 30 km, and checking 106 sewer lines. Additionally, to check the rising water levels in real time and prevent flood damage in advance, visual inspection, remote control, and automatic recording system were installed for rainwater pumping stations. The major change was made in 2007 when the storm water pipe design standard changed from 75 mm/h to 95 mm/h. This upgrade was motivated by the devastating experience in New Orleans from Hurricane Katrina in 2005. The Seoul government invested more than 0.7 billion to upgrade the storm infrastructure. Many district offices also expanded the capacity of storm drains to cope with the recurring flood. However, such a structural approach was not sufficient to reduce flood risk in subsequent years.

After the 2011 flood that inundated the City Hall, the Seoul government built flood prevention structures to adapt to climate change. The plan, which invests 4 billion US dollars for the next 10 years, establishes a drainage improvement plan in 34 major flood-vulnerable regions and adds rainwater pumping stations. In 2016, for the first time in history, realizing the limitation of the structural measures, the city added nonstructural measures such as communication and education to reach out to vulnerable communities. This plan identifies 240 districts around the city that are prone to different types of flooding (fluvial flooding, pluvial flooding, soil fillup, and landslide flooding), acknowledging various needs for different types of flooding. Even though the 2012 revised construction law does not allow the construction of new buildings in flood-prone areas, more than 40,000 houses have been constructed since then, indicating the failure of coordinating housing projects with flood risk reduction strategies. Additionally, floods are recurring problems in neighborhoods where elderly and low-income people live (Son and Ban 2022).

While some improvements have been made, such efforts did not prevent another flooding in August 2022. For example, during the August 2022 flood, the Shinwol underground tunnel to store stormwater was designed to hold 100 mm/h rainfall; it already operated with 18-53 mm/h rainfall, holding 70% of flood water capacity (Joongangilbo 2022). Like the Gangdong district, the root cause of the chronic flooding is that many residential and commercial buildings were developed in the lowlands. Once used for rice cultivation and other vegetables, the Gangnam district is more than 10 m lower than the surrounding region. To make matters worse, existing underground storm waterways were ill-designed, not allowing rainwater to drain quickly. Additionally, floodwater is collected into the upper section of a tributary of the Han River, creating floodwater concentrated in a small geographical area. While the construction is underway, it is still unknown if the upgraded system can handle extreme rainfall events, as they have become more common in recent years.

## 2.5 Recent Strategies for Reducing Flood Risk

Several flood risk reduction strategies have been implemented or suggested at various scales by different actors. Once a more proactive flood defense system is in place, flood risk and damage could be lowered. As a technology-savvy society with the fastest internet system in the world, emergency management agencies have actively used social media such as Facebook, Twitter, and YouTube to communicate flood disaster information. These social media uses were effective in creating positive images of these organizations as well as providing positive community emotional responses by facilitating the participation of citizens in risk communication (Kim et al. 2016; Song et al. 2015).

At a building scale, for example, one commercial building in the Gangnam district constructed a floodgate surrounding the building where the gate is closed when the water level rises to a critical level. The gate effectively defended flooding during most recent flood events while surrounding buildings were inundated. While a floodgate has effectively reduced flood risk at a building scale, district-wide efforts are yet to be implemented. If not planned at a district level, such a building-scale project could externalize flood risk to other surrounding places where such technology is not adopted, typically the people with fewer resources and power. Similarly, urban regeneration projects such as regenerative industrial compounds in Seoul could have decentralized stormwater retention and infiltration facilities to improve neighborhood sustainability and promote spatial quality (Hwang et al. 2020). As such, making space for water in public places through institutional arrangements must be made when devising flood risk reduction strategies.

Another strategy to reduce flood risk is through stream restoration at the district level. According to a recent study, Seoul's flood risk management has shifted from a purely technological approach to embracing social and ecological principles (Chang et al. 2021). A couple of examples are the increasing number of tributary stream restoration projects and the installation of green infrastructure. After the symbolic Cheonggye stream restoration, many local districts started to restore urban streams by opening culverts or re-meandering streams, providing space for recreation during low flow periods and absorbing water during storm events. Many citizen organizations have been active participants in these urban stream restoration projects, codesigning the desired outcome of the projects. The new flood risk management strategies also embrace a social approach to better inform its residents about the risk of flooding, particularly for those who live in half-basement. City officials have been actively reaching out to those vulnerable people to ensure their space is protected when urban street flooding happens. Additionally, green stormwater infrastructure is gaining increasing attention as a potential climate adaptation strategy for reducing stormwater runoff in Seoul (Lee et al. 2021). Researchers from Korea University, Seoul National University, and Sejong University used the Stormwater Runoff Reduction Module (SRRM) in a system dynamics model to evaluate the effectiveness of green stormwater infrastructure in reducing peak runoff amount and delaying peak runoff timing. By focusing on three districts that were not included in Seoul's urban greening plan, their study showed storm runoff got delayed by more than three hours and peak discharge amount got reduced by more than 20% in all three districts when green roofs, infiltration storage facilities, and porous pavement are installed (Song et al. 2022).

Large-scale sectoral or inter-district collaboration is also being recognized as a potential strategy to reduce flood risk. Because many rivers or streams cross jurisdictional boundaries, it is imperative to better coordinate river management policies among different districts and regulatory agencies. In South Korea, the central government manages major national rivers, maintaining levees along the major rivers and regulating flow releases from major upstream dams. The local government manages tributaries of the main rivers with different standards. While levees are designed to stand 200-year flood events for the major national rivers, they are designed to hold water from 50- to 200-year flood events for tributaries. With these different standards, local governments are not well prepared to cope with extreme flood events, potentially becoming victims of flooding or transferring flood risk further downstream. Tight coordination with different jurisdictions with standardized laws could help create a flood-resilient city. Shafique and Kim (2018) echoed that the lack of collaboration between different agencies could impede the expansion of low-impact development and green infrastructure in South Korea.

There are some promising signs of improvement in flood risk management in Seoul. By interviewing key flood practitioners in two districts' of Seoul, Ro and Garfin (2023) showed that institutional adaptive capacity to manage flood risk has increased over time. Like Chang et al. (2021), their findings show an emphasis on social (e.g., neighborhood outreach) and technological approaches (e.g., increasing flood risk monitoring). The same study also identified room for improvement in current practices. As emphasized by Chang et al. (2021), double-loop social learning processes are key ingredients for achieving flood resilience. Thus, a long-term risk management plan needs to be flexible enough to empower citizens to act and make necessary changes.

## 2.6 Conclusions

As one of the most affluent cities in the world, Seoul has a unique history of development and climate challenges. The city has a long history of managing excessive water during the monsoon season, dating back to the Yi Dynasty, by measuring precipitation and dredging streams to prevent overflow of water to surrounding regions. During the latter half of the twentieth century, the uncoordinated land and water planning exacerbated flood risks in many newly developed areas. To reduce flood risk, the city has traditionally focused on structural measures such as levees, storm pipes, and other hard infrastructure. Realizing the limitations of relying solely on these technological approaches, the city has now embraced more social and ecological principles to reduce flood risk, such as introducing green infrastructure and citizen education. While technology-based solutions still dominate flood risk reduction, highly educated and environmentally conscientious residents request more engaged management for reducing flood risks while maintaining environmental quality. Similar to Melbourne, however, the more affluent and privileged were able to more proactively engage with these dialogues to build such environmental organizations. Comprehensive flood risk management also requires intersectoral and inter-jurisdictional collaboration to remove institutional barriers. Climate change is likely to increase the probability of future flooding in Seoul, particularly toward the end of the twenty-first century. While there is high uncertainty in projecting future precipitation, temperature is projected to increase consistently among different global climate models. A simulation study at a subwatershed level in Seoul shows (Kim and Kang 2020) substantial increases in flooding under both low (RCP 4.5) and high (RCP 8.5) scenarios. However, when green infrastructure, such as water tanks, permeable pavement, and ecological waterways, was installed, flooding was projected to decrease substantially (Choi et al., 2021), indicating that proactive flood management could create a more flood-resilient city.

## References


Bae S, Chang H (2019) Urbanization and floods in the Seoul metropolitan area of South Korea: what old maps tell us. Int J Disaster Risk Reduct 37:101186

Chang H (2008) Spatial analysis of water quality trends in the Han River basin South Korea. Water Res 42(13):3285–3304. https://doi.org/10.1016/j.watres.2008.04.006

Chang H, Eom S, Yasuyo M, Bae D (2020) Land use change, extreme precipitation events, and flood damage in South Korea: a spatial approach. J Extreme Events 7(3):2150001

Chang H, Yu D, Markolf S, Hong C, Eom S, Song W, Bae D (2021) Understanding urban flood resilience in the Anthropocene: a social-ecological-technological systems (SETS) learning framework. Ann Am Assoc Geogr 111(3):837–857

Choi G (2016) Changes in means and extreme events of Changma-period precipitation since mid-Joseon Dynasty in Seoul, Korea. J Korean Geogr Soc 51(1):23–40

Choi Y, Kang J, Kim J (2021) Urban flood adaptation planning for local governments: hydrology analysis and optimization. Int J Disaster Risk Reduct 59:102213

Hwang K, Schuetze T, Amoruso FM (2020) Flood resilient and sustainable urban regeneration using the example of an industrial compound conversion in Seoul, South Korea. Sustainability 12(3):918

JoongAngIlbo (2022) August extreme rainfall in Seoul – once in 500 years? – different from European floods. https://www.joongang.co.kr/article/25118744#home. Accessed 10 Sept 2023

Jung GS (1990) Lifeline but dangerous river. Sisa J. https://www.sisajournal.com/news/articleView.html?idxno=112468. Accessed 12 Sept 2023

Kim J, Kang J (2020) Analysis of flood damage in the Seoul Metropolitan government using climate change scenarios and mitigation technologies. Sustainability 13(1):105

Kim K, Jung K, Chilton K (2016) Strategies of social media use in disaster management: lessons in resilience from Seoul, South Korea. Int J Emerg Serv 5(2):110–125

Kim HR, Moon M, Yun J et al (2023) Trends and Spatio-temporal variability of summer mean and extreme precipitation across South Korea for 1973–2022. Asia-Pac J Atmos Sci 59:385–398. https://doi.org/10.1007/s13143-023-00323-7

Korea Meteorological Administration (2023) Station meteorological data. National Climate Data. https://data.kma.go.kr/climate/RankState/selectRankStatisticsDivisionList.do

Korea Ministry of Interior and Safety (2022) 2021 Statistical yearbook of natural disaster. Sejong. https://www.safekorea.go.kr

Lee JY, Kwon KD, Raza M (2018) Current water uses, related risks, and management options for Seoul megacity, Korea. Environ Earth Sci 77:1–20

Lee H, Song K, Kim G, Chon J (2021) Flood-adaptive green infrastructure planning for urban resilience. Landsc Ecol Eng 17:427–437

Ro B, Garfin G (2023) Building urban flood resilience through institutional adaptive capacity: a case study of Seoul South Korea. Int J Disaster Risk Reduct 85:103474. https://doi.org/10.1016/j.ijdrr.2022.103474

Seoul Research Institute (2013) Geographical atlas of Seoul. Seoul

Shafique M, Kim R (2018) Recent progress in low-impact development in South Korea: water-management policies, challenges and opportunities. Water 10(4):435

Son CH, Ban YU (2022) Flood vulnerability characteristics considering environmental justice and urban disaster prevention plan in Seoul Korea. Nat Hazards 114(3):3185–3204. https://doi.org/10.1007/s11069-022-05511-8

Song M, Kim JW, Kim Y, Jung K (2015) Does the provision of emergency information on social media facilitate citizen participation during a disaster? Int J Emerg Manag 11(3):224–239

Song K, Kim M, Kang H-M, Ham E-K, Noh J, Khim JS, Chon J (2022) Stormwater runoff reduction simulation model for urban flood restoration in coastal area. Nat Hazards 114(3):2509–2526. https://doi.org/10.1007/s11069-022-05477-7

# Chapter 3
# Jakarta, Indonesia

**Abstract** The site of Jakarta was developed by colonial powers and involved the structural dilemmas of flooding and disease from the start. The problems of flooding and water quality are shown to coincide with the dispersal of informal settlements, which are subsequently blamed for the contamination of water and displaced to other vulnerable areas, while developers replace them with luxury buildings. The faulty planning, extending from the shaky foundations of colonial inequality through the repressive New Order of President Suharto, sustained an unsustainable process of hyper-urbanization without critical foresight regarding water hazards renders the megacity's adaptive capacity incapable of confronting the problems of sea level rise, pollution, and flooding. While there are some efforts to stem the tide of climate hazards, from massive projects such as building a sea wall to smaller, localized efforts to recharge aquifers with rainwater, the pace of climate change appears to outstrip the city's ability to establish viable solutions for its enormous and growing population. For this reason, Jakarta is the first and, so far, only city in our assessment to begin the process of "managed retreat" to a different island with the intention of decreasing population pressures and planning means to promote a more sustainable future.

**Keywords** Megacity · Seawall · Subsidence · Informal settlements · Managed retreat

## 3.1 Introduction

Water challenges are nothing new to Jakarta, but climate change is deepening them. The capitol of Indonesia, Jakarta, must contend with a legacy of maladaptive growth inherited from colonialism and its institutional influence on the postcolonial New Order government from the mid-1960s until President Suharto's ouster in 1998. Because they emerged through the convergence of social, economic, and political causes exacerbated by climate change, Jakarta's water issues require complex solutions that integrate different sectors into a holistic approach to equity in water management.



H. Chang, A. R. Ross, *Climate Change, Urbanization, and Water Resources*, https://doi.org/10.1007/978-3-031-49631-8_3

Jakarta's physical geography lends itself to fluvial flood patterns, but development caused natural tendencies to intensify (Tambunan 2017). The developmentalist state of postcolonial Indonesia organized a water management regime that improved, but did not significantly alter, the trends of colonial power (Kooy and Bakker 2008a). With the New Order system, the Indonesian government's water management system reified class disparities in water management. Migrants moving to the city built informal developments outside water service providers, and new impervious surfaces replaced green spaces along riverbanks (Rukmana 2015).

The result of displacement and new development is a pattern of fluvial flood intensification, through which the poor suffer the most. At the same time, a lack of services to the poor contributes to water pollution and the informal extraction of subsurface water, which contributes to the subsidence of Jakarta's delta, thus causing more risk of tidal floods and increasing the precarity of informal settlements in those areas (Bakr 2015). In order to adjust water management to meet these multifarious challenges, the City of Jakarta and the government of Indonesia have worked to advance large-scale, ultramodern engineering projects ranging from infrastructure improvements to a massive seawall to an artificial island metropolis (Octavianti and Charles 2018). Yet, these efforts to reduce flood risk do not address the systemic problems of modernist development and land use, reinforcing class disparities at the heart of the postcolonial crisis.

We argue that a heavier focus on land use, along with a commitment to greater equity within metropolitan Jakarta, would significantly reduce fluvial flooding and promote the sustainable development of the megacity as a model of adaptive transformation to the dual challenges of climate change and structural causes of hydrological crises. While Jakarta is taking some necessary steps toward these goals, more sustainable initiatives could be implemented to continue the movement toward equitable water systems. The city appears to be at an impasse, negotiating a managed retreat to a new capitol city while attempting to address climate issues at the same time. If the retreat is successful, Jakarta will have been the first megacity downgraded by climate change.

## 3.2 Climate and Physical Geography

Jakarta lies in a fertile delta on the northwest coast of the long and narrow island of Java in the Indonesian archipelago formed by volcanic eruptions that characterize the "Ring of Fire" around the Pacific Ocean. More specifically, Java and neighboring Sumatra jutted from the ocean through a tectonic process caused by the grinding and overlapping, or "subduction", of the India and Australian plates. From this subduction, pressure and steam cause lava to spew from the Java Trench under the ocean, forming volcanos and islands in a pattern known as the Sundra Arc (Whittaker et al. 2007). Jakarta sits at a strategic point near Sumatra, offering an important wayfaring and defensive position for Java's inland inhabitants. Beyond the island of

Sumatra, to the west lies Singapore and the mainland of Southeast Asia. To the northeast lies Kalimantan, with Sulawesi further across the Java Sea.

Like most great cities, much of Jakarta's history can be told by those rivers that shaped its geography. Looking south from the megacity on a particularly clear day, one can just make out the formidable green slopes of Mount Pangrango, the origin of the Ciliwung River, which winds its way 119 km back through Jakarta and out to the Java Sea. While the Ciliwung transports Pangrango's lush, volcanic soils along its narrow route through the mountains, depositing them in the delta, it also provides a critical resource for upstream habitat. The mountains and their surrounding rainforests play host to hundreds of bird species, along with the charismatic silvery gibbon, the Javan leopard, and the "barking deer" (Whitten et al. 1996) (Fig. 3.1).

The names bestowed on these features by the native Sundanese people tell a story in themselves. Pangrango means "that which huffs and puffs," revealing the locals' relationship to the volcano that towers 3019 m above sea level. The river's name, Ciliwung, likely derives from a common word shared by the Sundanese and Javanese inhabitants of the area, *liwung*, which means distressed, twisted, or

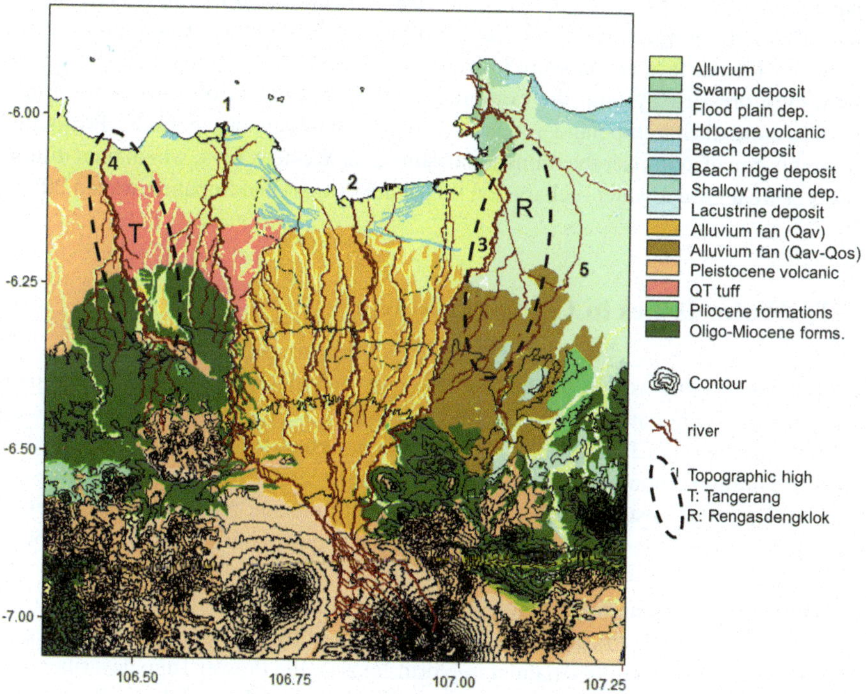


**Fig. 3.1** Elevation, drainage patterns, and geology in Jarkarta. The funnel-shaped basin is "bounded by the Cisadane (1) and Kali Bekasi (3) Rivers in the west and east, respectively." The Cisadane (1), Ciliwung (2) and Kali Bekasi (3) Rivers originated from surrounding mountains and form "the narrow base of a funnel-shaped topographic low with Jakarta Bay at its mouth. The Cimanceuri River is labeled as (4). Tangerang and Rengasdengklok Highs are indicated by dashed lines, labeled T and R, respectively." (Figure source: Cipta et al. 2018)

troubled—a testimony to its turbidity and frequent floods and inundations caused by heavy rains (Grijns 2001). The uncontacted, ethnically Sundanese Baduy Kanekes people fastidiously maintain ancient shrines to the creation of the world that they believe happened in these foothills. And it might be said, within the complexity of this dense ecosystem, universes exist.

Pollen records provide scientists with an interesting and complex understanding of paleoclimatological records in the area beginning around 140,000 years ago, when the climate of the Bandung delta was characterized by strong seasonality and a yearly rainfall of 750–1000 mm. From 126,000 Before Present (BP), the regional climate saw some variability but remained more humid and wet until 81,000 BP, when precipitation likely fell back down to 1000 mm, the monsoons moved less moisture, and seasonal runoff increased. Subtle changes took place in the ensuing years, with some cooling possibly transpiring between 62 and 47,000 BP, followed by a dryer phase until 20,000 BP and significantly cooler temperatures during the last glacial period prior to the amelioration of the climate (Simanjuntak 2006).

While the island of Java has a very scanty forested area left, forests in other parts of the archipelago are dense in biodiversity. However, other islands such as Borneo have seen an increase in the average number of hectares on fire annually, particularly during prolonged droughts brought on by El Niño-Southern Oscillation. As the typically low surface air pressure moves over to the Central Pacific and is replaced by a high-pressure system, the evaporative moisture content of air over the warm pool around the archipelago reduces, bringing on long dry periods. Prolonged droughts can also impact the monsoonal rains over Western Java, which then diminishes critical groundwater recharge on the island (Supari and Sopaheluwakan 2016).

## 3.3 Water Issues in Colonial Batavia

Storied and mighty as it is, the Ciliwung is only one of 13 rivers offering their own unique story and contribution to Jakarta's prolific delta region. Water has made up both the lifeblood of Jakarta and its most pressing issues since the first native settlements. Through colonialism, water infrastructure created a racialized class division that ran into the postcolonial era. Today, climate change redoubles the systemic crises of water distribution, sanitation, and inequity born of colonial power dynamics.

Water plays such an important role in the life of the city that the oldest historical record of Jakarta, the Prasati Tugu, is a document of hydraulic projects. Written in Sanskrit and discovered in a hamlet in North Jakarta, the Prasati Tugu describes stories in Sanskrit a canal project diverting water from the Candrabhaga river past a royal palace in the delta of today's sprawling megacity. The inscription notes that the Brahmans responsible for the canal received 1000 cows in return, elucidating some of the class and labor relations during the period. Other irrigation and drainage projects took place during those ancient times to cultivate crops, manage flood risk, and produce monuments to the Sundanese Taruma Kingdom (Noorduyn and Verstappen 1972).

Some 1200 years later, in 1619, the Dutch founded the colonial City of Batavia on those grounds with the lofty title of the "Queen of the East." As a result of rampant waterborne diseases, however, Batavia earned a new nickname "Graveyard of the East" (Dewi et al. 2018). To escape the ill vapors, humors, and miasmas believed responsible for the *ziektenhaard*, or breeding ground for disease, the Dutch spread out in Swiss-style chalets around the countryside (Kooy and Bakker 2008a). Avoiding the delta helped the colonists develop the *cultuur* system, through which the authorities could directly control one-fifth of all plantation land. With the development of scientific knowledge about the health impacts of tainted water supplies in the mid-nineteenth century, though, the Council of Indies in Holland made plans to develop seven artesian wells supplying Batavia's 8000 residents with fresh water from hydration points (Kooy and Bakker 2005). This turn toward hydraulic development indicated a pivot whereby "the new purpose of the colonial government was to encourage the increased participation of the private sector, while maintaining the legitimacy of a system of colonial rule that enabled profits to European investors" (Kooy and Bakker 2008a).

Water supply came to define a boundary between the traditional systems of water testing created by natives who relied largely on rivers and streams, on one hand, and the colonial population, which increasingly identified itself through a separate regime of health and sanitation on the other. By the early twentieth century, planners abandoned the sulfur-infused artesian wells for extensive water-piping projects connecting homes in a "spiderweb-like" fashion. Basing the new developments on ethical principles to develop native populations suffering from diseases and "keep them out of the kalis (canals)," colonial authorities thus laid the groundwork for the uneven development of the colonial hydraulic regime seeking to draft Batavia on the model of a modern European city (Kooy and Bakker 2008b).

Development plans, themselves, called for unequal distribution of water serving 90% of the European households received pressurized water piped from springs and only 33% of the native population, many of whom relied on hydrants. Thus, Europeans consumed the majority of spring water in colonial Batavia right up until the end of the colonial period. Indeed, water infrastructure improvements increased property values in affected districts, leading to a further spatial differentiation between the haves and have-nots. This pattern of uneven development continued into the postcolonial period, as middle-class bureaucrats advanced into the better-serviced residential districts (Surbakti et al. 2010).

## 3.4 Postcolonial Inequality

With an eye to the modernization plans of President Sukarno's "city beautiful" projections for Jakarta, Kooy and Bakker write, "the piped network mirrors the above ground highways and flyovers built to connect the modern elements of the city, and was neatly positioned by the government to channel the increased flows of water to follow the new flows of international traffic into modern areas of the city, simultaneously excluding the vast majority of un-modern spaces and populations from both spatial proximity to, and services from, the network" (Kooy and Bakker 2015). Indeed, the postcolonial system was built on the foundations of the colonial one. Rather than successfully centralize and expand the water infrastructure that existed, the postcolonial system deepened vulnerabilities for the area's poor, which would become exacerbated by water issues posed by anthropogenic climate change.

As new monuments to international transit and trade erupted in developing regions of Jakarta, water treatment plants and pipe networks sketched out a delimited urban area outside of which non-modern residents lived according to traditions scorned by progress. Even the satellite towns created by the Dutch to avoid the miasma received piping and service, revealing the extent to which postcolonial infrastructure maintained course, despite the influx of migrants to open areas of Jakarta. The massive water projects of the postcolonial era drew largely from the 1957 Jatiluhur Reservoir, via the West Tarum Canal, but even those 8000 hectares could not meet the needs of Jakarta's growing population (Bahri 2020).

A postcolonial class division emerged in which the poor drew from open wells, surface water, and hydrants, on which the government placed a higher tariff than piped water. Also, the expansion of the urban area accelerated the encroachment of developed land on local agriculture in the southern areas. Conversion of land surfaces to impermeable sidewalks and streets increased the potency of area floods. As more people flowed in and the area modernized, rivers also became polluted by industrial actors and sewage, depriving residents of their traditional water source and also increasing the risk of flooding by adding toxics (Furlong and Kooy 2017). Forced to rely more on groundwater due to surface water pollution, urban residents produced informal settlements inclusive of added withdrawals from shallow aquifers, which can be polluted by toxics entering wells during floods (Bakker 2007).

Today, piped water covers 66% of the city, but just 800,000 households fall on the list of residents served by the state water company. The city produces more than 8000 tons of waste per day, properly treating only half of it (Aprilia 2021). The reason for this low service stems from both the unequal development of water and sanitation pursuant to the colonial regime and from privatization regimes set into place after the coup that replaced President Sukarno with the dictatorship of Suharto in 1966. The establishment of private companies to distribute water and provide sanitation services privileged formal housing developments whose residents had reliable purchase power (Kurniasih 2008).

Informal settlements that rely on irregular employment often resort to a hybrid system of centralized water supply and decentralized sources, such as cisterns, septic tanks, and bottled water. Where water providers do lay pipes, low pressure in the low-lying areas often results in intermittent supply, leading residents to rely alternately on ground wells and pipes. On the other hand, where higher quality service is set into place, often as a result of area land grabs and high-income developments, informal settlements are priced out of the water market completely (Kurniasih 2008).

## 3.5 Development of Flooding

Suharto's regime fastened the national budget tightly to the Organization of Petroleum Exporting Counties (OPEC) and oil exports, such that the volatile oil markets of the 1980s, added to the projected decline in oil supply and decrease in crop prices, led to an immediate need to diversify the economy. The New Order regime opened Jakarta up to direct investment and attractive megaprojects for consumers, including housing developments and shopping malls. As the city grew into a megacity, new export-processing zones grew, the city developed an international airport, and new business services emerged. The privatization and deregulation of state sectors such as water management and land use would later bring about significant crises (Bakker 2010).

People started to pour into Jakarta to get a piece of its expanding economy, redoubled as the increase of large agricultural and industrial development in the hinterland brought about displacement to the cities. While Suharto's expansion of extractive industries in rural Java pushed residents into the city at a rapid rate, land grabbing from informal areas in more valuable real estate within the metro area forced poor residents into areas more prone to flooding. Due to urbanization practices, population growth, and climate change, studies have ranked Central and Northern Jakarta as the first and second most vulnerable places in Southeast Asia to the effects of climate change (Firman et al. 2011), respectively. The impacts of sea level rise and flood intensity are worsened by urbanization, population growth, and land use.

Low service and high levels of groundwater pumping caused the land to subside, thus becoming more flood prone. A large flood in 1996 portended significant shocks to come. The next year, the global economy collapsed, and Jakarta was hit particularly hard. As a result of the crisis, Suharto stepped down. Yet his legacy remained, as land management practices continued to escalate periodic extreme flood events.

The share of undeveloped areas within the basin of the Ciliwung River shrank from 66% in 1970 to just 38% in 2000 as development came to dominate the region, leading to greater flooding issues (Fachrul et al. 2007). The crisis of accelerated urban development had been identified by the Indonesian government by 1992, with the passage of the first spatial planning law, but its novel status and insufficiently delineated regulations failed to meet the most pressing problems. By 2008, virtually all green space had vanished. Deforestation, along with garbage from informal settlements without access to city infrastructure, can lead to more rapid flooding and clogging of waterways. Because it is the poor who tend to be blamed for blocking waterways, flooding becomes an excuse to displace the most vulnerable people to even more vulnerable areas while increasing urban development (Dovey et al. 2019). The problem is pressing, as the crisis of subsidence increases throughout Jakarta, and is worsened by rising sea levels, due to melting icecaps caused by anthropogenic climate change.

While Jakarta saw extensive flooding in 1996 and 2002, the worst floods to hit the capitol came in February 2007. Impacting some 60% of the urban area, the floods killed some 58–74 people, affecting around 400,000. The banks of the Ciliwung topped 11.2 m above the thalweg, and in informal settlements, a hundred houses met with complete destruction. While the intensity of the storm surpassed that of the flood of 2002, the two dropped a fairly equal amount of rainwater. According to scholar Pauline Texier, the magnitude of the 2007 flood can be explained by "anthropogenic factors which enhance the natural flooding process of a meteorological or hydrological nature (monsoonal climate, subsidence and topography, rivers network)" (Texier 2008).

As climate change intensifies, flooding will only worsen, but human efforts can mitigate the impacts (Mishra et al. 2018). According to the Meteorology, Climatology, and General Physics, not only will climate change cause storms to intensify as a result of increased sea level and the quantity of water vapor in the atmosphere, but it is also already leading to increased temperatures in Jakarta. While some islands in the Indonesian archipelago have experienced a 1 °C temperature rise, Jakarta experienced a 0.82 °C rise in 2019 over the 2000–2010 levels. From 1866 to 2010, Jakarta's temperature increased at nearly twice the rate of global warming, indicating that the city is perhaps more vulnerable than most (Cahya 2020).

In the aftermath of another terrible flood in 2013, a team of scientists led by Yus Budiyono at Jakarta's Agency for the Assessment and Application of Technology developed a flood risk assessment model. Adapting the open-source Python toolkit, DamageScanner, Budiyono's team found annual expected costs of $321 million per year as a result of floods. Indeed, as a result of sea level rise and subsidence, by 2050, areas of North Jakarta may be completely inundated (Budiyono et al. 2016).

## 3.6 Climate Risks

Flooding is only one increased environmental risk among the myriad challenges presented by climate change to the island of Java and more directly to the City of Jakarta. Climate change will likely change not only the intensity of weather events but also their timing, leading to some drier seasons and some wetter seasons. This phenomenon will likely draw more fluvial flooding in some cases and will also likely increase the potential for forest fires in others—particularly during El Niño events. These trends will likely impact water availability, crop growth, and water quality on the island (Pawitan et al. 1999).

Pollen records provide scientists with an interesting and complex understanding of paleoclimatological records in the area beginning around 140,000 years ago when the climate of the Bandung delta was characterized by strong seasonality and a yearly rainfall of 750–1000 mm. From 126,000 BP, the regional climate saw some variability but remained more humid and wet until 81,000 BP, when precipitation likely fell back down to 1000 mm, the monsoons moved less moisture, and seasonal runoff increased. Subtle changes took place in the ensuing years, with some cooling possibly transpiring between 62 and 47,000 BP, followed by a dryer phase until 20,000 BP and significantly cooler temperatures during the last glacial period prior to the amelioration of the climate (Simanjuntak 2006).

While the island of Java has very little forested area left, forests in other parts of the archipelago are dense in biodiversity. However, other islands such as Borneo have seen an increase in the average number of hectares on fire annually, particularly during prolonged droughts brought on by ENSO. As the typically low surface air pressure moves over to the Central Pacific and is replaced by a high-pressure system, the evaporative moisture content of air over the warm pool around the archipelago reduces, bringing on long dry periods. Prolonged droughts can also impact the monsoonal rains over Western Java, which then impacts critical groundwater recharge on the island (Supari and Sopaheluwakan 2016).

As climate change intensifies the trends in natural variability cycles, droughts in Indonesia and Australia lead to enhanced forest fire seasons, contributing to the largest sources of Indonesia's green house gas emissions. This feedback loop of climate change bringing on more fires, leading to more emissions, can prove disastrous for air quality in the area, as well as water quality impacted by smoke and ash carried from massive fires. Depending on the position of the intertropical convergence zone, the transport of smoke can move as far as the Indian Ocean, with sweeping consequences (Tapper 1999).

In addition, the increase in carbon emissions causes acidification of the ocean, as the carbon in the atmosphere reacts with water molecules, producing carbonic acid, which then breaks down into bicarbonate and hydrogen ions that lower the ocean's pH. Through that process, coral reefs around the Indonesian archipelago die off, and marine life that would otherwise provide the ocean's filtering system are unable to produce and maintain their shells. Again, a feedback mechanism of aggressive climate change destroys not just the ecosystem but its capacity to cope with that destruction (Lam et al. 2019).

The flip side of dry periods is periods of prolonged rainy periods, and more intense rainy seasons, which can contribute to more problems for Jakarta. Central and North Jakarta are the two most vulnerable places in Southeast Asia, due to flooding. As melting ice sheets and glaciers increase sea levels around the world, the current rate of warming would see most of Central Jakarta inundated. Disease outbreaks would also likely become more common as a result of increased precipitation incidents (Firman et al. 2011).

Also, riverine water quality will be negatively affected by climate change. The rates of deterioration stand to increase for $NO_3$, dissolved oxygen, and chemical oxygen demand, making the water more contaminated (Kumar et al. 2017). Furthermore, droughts lead to more need for reused water, increasing the problem of contamination (Wijaya and Soedjono 2018). This will likely have the largest effect on poorer communities who rely on rivers for more environmental services.

## 3.7 Spatial Planning

The experience of the increasing flood intensity helped shape Jakarta's transition to democracy after three decades of authoritarian rule under Suharto's "New Order." Though the 1945 Constitution asserted the government as the highest power of political control, planning regimes implemented in 1992 and reemerging in 2005 favor spatial coordination over economic development, decentering the state while also ensuring that plans exist to limit the scope of haphazard settlement. While this mode of operation helps include citizens at the level of consultation, it clashes with technocratic traditions in the postcolonial bureaucracy, creating a new political culture of contention over the normativity of extant traditions.

Specifically, the amended Spatial Planning Act (Law No. 26/2007) mandated that 30% of urban lands must be left for open space. The act also established the Jabodetabekpunjur spatial plan for the entire metropolitan area, encouraging a decentralized process through which each area can develop its own spatial plan modeled on integrated development. Within this framework, individual spatial plans are to take into account soils, environmental capacity, and water quality and quantity. Also, the Bogor Regency, a crucial metropolitan water source, is demarcated for environmental conservation (Rukmana 2015).

By empowering autonomous local forms of government over central planning, the new spatial regime is meant to increase democracy but in cooperation with private sector powers that may undermine the prioritization of spatial coordination in favor of profit rather than economic development, per se. Hence, a biopolitical order meant to destabilize hierarchical concepts of economic development in order to support democratic organization can have the opposite effect of using the auspices of democratic organization to promote corporate profit over economic development (Wade 2020). In this regard, spatial planning can reinforce inequality despite intending to expand access to collaborative power if sufficient protections are not in place to ensure the economic security of local populations.

In the case of water management, inequality can arise through privatization schemes that do not protect against rate hikes, displacement through development schemes, and prejudicial access to services. A World Bank loan of $500 million required the privatization of water management, subsequently enshrined into law with the 2004 Water Resources Act. As the 1945 Constitution designates water as a basic need and rejects full privatization, the 2004 Water Resources Act faced criticism for its lack of clarity in the extent of private involvement in water management allowed (Kurniasih 2008). Thus, the new policy was viewed as a capitulation of public authority to the private sector based on the failure of postcolonial administrations to successfully mitigate water contamination, scarcity, and natural disasters. This capitulation proved dysfunctional to its critics precisely because of the ambiguity it created between the provinces of private and public sectors without clearly producing partnership between them (Leong 2015).

In their article, Padawangi and Douglass (2015) argue for a solution that unites "the micro-geographies of community formations in flood-risk areas and the city region scale of relations between land use changes and the urban ecology," stating "actions at either scale must coordinate with the other if sustainable solutions are to be found." As neoliberal reforms reduced government spending on environmental protections, localized community formations produced new discourses and practices of household water management (Putri and Moulaert 2017). Yet, this pattern only reproduces systems of displacement to informal settlements increasingly prone to flooding and pollution.

Following the catastrophic flood events, the government developed a National Action Plan for Disaster Risk Reduction 2010–2012. From the National Action Plan, a number of reasonable proposals emerged. Firstly, a coastal project to defend against flooding was produced with the help of the Netherlands. Also, the government gained World Bank support in creating an emergency dredging initiative to better manage riverine flooding. Another comprehensive flood management program promised to upgrade water infrastructure, and lastly, an inter-ministerial committee called the Committee for Acceleration of Priority Infrastructure Delivery proposed a massive sea wall (Padawangi and Douglass 2015). Rather than force the sea wall project through, however, the city disincentivizes other possible initiatives, ensuring that criticisms of the failure to address the causes of subsidence would be subsumed in the scientific process (Octavianti and Charles 2018).

While these plans addressed inadequate storm drainage infrastructure, they did not confront the more structural problem of land use change, despite its critical impacts on flood intensification. Moreover, officials used flood management as an excuse to relocate 350,000 people from the Ciliwung riverbanks. Also, developers claim to combat the rising sea levels and subsiding land of Jakarta by engaging in a large-scale land reclamation projects for golf courses and the expansion of the Taman Impian Jaya Ancol amusement park, as well as a free trade "special economic zone" for manufacturers and Pluit City, a proposed engineered island off the coast of Jakarta—"an amazing modern city that would be a plus for the capital City of Jakarta" (Podomoro 2014).

While spatial planning appears to be on the agenda for Indonesia, Jakarta's efforts thus far have largely existed in terms of either functional infrastructure projects or grand schemes inclusive of the creation of a new island city. What policymakers appear to have avoided is the effects of urbanization, itself, and especially the creation of impervious surfaces where green spaces once absorbed much of the flood waters (Rahardjo 2017). To confront this gap in policy, more efforts are emerging to develop green infrastructure in the city and promote the extension of water infrastructure to the needy.

## 3.8 Adaptation

With only 7.02 $m^2$ of green space for every person, Jakarta offers far less in the way of green infrastructure than cities such as Stockolm (80 $m^2$), New York (30 $m^2$), or Paris (15 $m^2$) (Kirmanto et al. 2012). This paucity of green spaces comes at the expense of air quality, while intensifying the heat island effect and impacting Jakarta's water system. With more green space, aquifers will recharge at a greater rate, thus compensating for some of the challenges of subsidence, and the risk of flood damage will decrease.

In 2011, Indonesia shifted planning imperatives toward "livability" through the Green City Development Program developed by the Ministry of Public Works. This Green City program focuses on a gradual transition toward increasing green energy, transportation, space, community, planning and design, and building, and green waste and green water. Jakarta has engaged voluntarily with numerous other municipalities and regencies to promote active public participation in green infrastructure, i.e., green communities. By utilizing school visits and public lectures, along with social media activity and advertisements in traditional media, the Green City Development Program is beginning to increase focus on multiscalar approaches to expanding biodiversity and green spaces both large and small. Different municipalities participate in the "green map," an active document for urban planners, designers, and architects to identify opportunities for smart, green infrastructure from the neighborhood scale to the urban and periurban scales (Kirmanto et al. 2012).

Jakarta's peri-urban areas have also become a focus of attention for sustainable water management. Urban sprawling around Jakarta, in what is known as the Jabodetabek metropolitan area, can extend 25 km out from the city limits. These areas built by private developers as "dormitory towns" have an opportunity to develop in keeping with standards that afford greater green open spaces to residents (Hasibuan et al. 2014).

## 3.9 Conclusions

Jakarta benefits from its position in a lush, rainforest landscape that serves as a hotspot for biodiversity served by multiple water sources. The city's development tamed to a great extent the prevalence of deadly diseases and the crises of fetid water quality. However, urbanization also brought with it large-scale destruction of the landscape, partly to control the problems of disease. Today, the strength of social, ecological, and technological system approaches shows that further integration between the ecological and urban areas could contribute to problem-solving by way of flood mitigation.

Among the greatest challenges facing Jakarta lies in the unequal distribution of flood impacts, and the feedback effect through which they render the unsustainable precarity of informal domiciles more permanent (Padawangi 2012). Rather than reinforcing the problem of income insecurity and systemic poverty by focusing on clearing away the poor from informal settlements, while focusing on techno-solutions for the urban rich, Jakarta could put more resources toward establishing housing and sustainable economic development for the poor through adaptive water policies and planning (Dwirahmadi et al. 2019). Improving flood mitigation based on efforts attempted in other areas of Indonesia, such as the creation of biopores to improve rainwater infiltration in Bandung and Malang, could encourage less surface runoff while fending off subsidence that causes worse flooding (Drosou et al. 2019). Meanwhile, increasing rainwater harvesting in private residences can take some of the burden off of the sewer systems.

However, authorities in Jakarta appear to have decided that it is too late for Jakarta. Plans to relocate the capitol of Indonesia are already in advanced stages. The site lies some 620 miles away in East Kalimantan at Nusantara, and the move is slated to cost $32.5 billion. More central and less prone to natural disaster, Nusantara spreads out across 180,000 hectares, housing 1.5 million civil servants on jungle terrain currently home to orangutans and other species (Al Jazeera 2022a, b).

The case of Jakarta's battle with climate change-induced water hazards does not inspire much in the way of hope, but it does provide an example of a strategy less discussed in the context of this volume: managed retreat. According to A.R. Siders, managed retreat involves "the purposeful, coordinated movement of people and assets out of harm's way" (Siders 2019). Although it obviously enjoys less enthusiastic support among policymakers, due to the appearance of giving up and the difficulty of displacement, scholars suggest that managed retreat increasingly presents the best solution in some areas. Historians and the lay public view mass displacement caused by past infrastructural development with regret and sadness at the misfortune of those impacted; however, today, controlled climate adaptation may offer a more humanitarian pathway in some areas (Carey 2020).

## References


Al Jazeera (2022a) Why Indonesia is abandoning its capital city to save it. https://www.aljazeera.com/news/2022/11/9/hldwhyindonesia-is-abandoning-its-capital-jakarta-to-save-ithld. 9 Nov 2022

Al Jazeera (2022b) Indonesia names new capital Nusantara as MPs back relocation. https://www.aljazeera.com/news/2022/1/18/nusantara-indonesias-parliament-passes-law-to-relocate-capital. 18 Jan 2022

Aprilia A (2021) Waste Management in Indonesia and Jakarta: challenges and way forward. In: Proceedings of the 23rd ASEF Summer University, Virtual, pp 1–18

Bahri M (2020) Analysis of the water, energy, food and land nexus using the system archetypes: a case study in the Jatiluhur reservoir, West Java, Indonesia. Sci Total Environ 716:137025

Bakker K (2007) Trickle down? Private sector participation and the pro-poor water supply debate in Jakarta, Indonesia. Geoforum 38(5):855–868

Bakker K (2010) Privatizing water: governance failure and the world's urban water crisis. Cornell University Press, Ithaca

Bakr M (2015) Influence of groundwater management on land subsidence in deltas: a case study of Jakarta (Indonesia). Water Resour Manag 29(5):1541–1555

Budiyono Y, Aerts JC, Tollenaar D, Ward PJ (2016) River flood risk in Jakarta under scenarios of future change. Nat Hazards Earth Syst Sci 16(3):757–774

Cahya GH (2020) Climate change cause of Greater Jakarta floods, BMKG says. The Jakarta Post

Carey J (2020) Managed retreat increasingly seen as necessary in response to climate change's fury. Proc Natl Acad Sci 117(24):13182–13185

Cipta A, Cummins P, Irsyam M, Hidayati S (2018) Basin resonance and seismic Hazard in Jakarta Indonesia. Geosciences 8(4):128. https://doi.org/10.3390/geosciences8040128
Dewi EP, Kurniawan KR, Ellisa E (2018) Transformation of canals in colonial Batavia. Int J Archit Urban Stud 3(1):53–63
Dovey K, Cook B, Achmadi A (2019) Contested riverscapes in Jakarta: flooding, forced eviction and urban image. Space Polity 23(3):265–282
Drosou N, Soetanto R, Hermawan F, Chmutina K, Bosher L, Hatmoko JUD (2019) Key factors influencing wider adoption of blue–green infrastructure in developing cities. Water 11(6):1234
Dwirahmadi F, Rutherford S, Phung D, Chu C (2019) Understanding the operational concept of a flood-resilient urban community in Jakarta, Indonesia, from the perspectives of disaster risk reduction, climate change adaptation, and development agencies. Int J Environ Res Public Health 16(20):3993
Fachrul MF, Hendrawan D, Sitawati A (2007) Land use and water quality relationships in The Ciliwung River Basin Indonesia. In: International Congress River Basin Management
Firman T, Surbakti IM, Idroes IC, Simarmata HA (2011) Potential climate-change related vulnerabilities in Jakarta: challenges and current status. Habitat Int 35(2):372–378
Furlong K, Kooy M (2017) Worlding water supply: thinking beyond the network in Jakarta. Int J Urban Reg Res 41(6):888–903
Grijns K (2001) JABOTABEK place names. In: Jakarta Batavia. Brill, pp 211–227
Hasibuan HS, Soemardi TP, Koestoer R, Moersidik S (2014) The role of transit oriented development in constructing urban environment sustainability, the case of Jabodetabek, Indonesia. Procedia Environ Sci 20:622–631
Kirmanto D, Ernawi IS, Djakapermana RD (2012) Indonesia green city development program: an urban reform. In: 48th ISOCARP Congress, vol. 4
Kooy M, Bakker K (2005) Splintered networks? Water, power, and knowledge in Jakarta 1870–1945. In: Workshop placing splintering urbanism: changing network service provision and urban dynamics in cross-national perspective, pp 22–24
Kooy M, Bakker K (2008a) Splintered networks: the colonial and contemporary waters of Jakarta. Geoforum 39(6):1843–1858
Kooy M, Bakker K (2008b) Technologies of government: constituting subjectivities, spaces, and infrastructures in colonial and contemporary Jakarta. Int J Urban Reg Res 32(2):375–391
Kooy M, Bakker K (2015) (Post) colonial pipes: urban water supply in colonial and contemporary Jakarta. In: Cars, conduits, and kampongs. Brill, pp 63–86
Kumar P, Masago Y, Mishra BK, Jalilov S, Emam AR, Kefi M, Fukushi K (2017) Current assessment and future outlook for water resources considering climate change and a population burst: a case study of Ciliwung River, Jakarta City, Indonesia. Water 9(6):410
Kurniasih H (2008) Water not for all: the consequences of water privatisation in Jakarta, Indonesia. In: Biennial conference of the Asian Studies Association of Australia. Melbourne, Australia
Lam VWY, Chavanich S, Djoundourian S, Dupont S, Gaill F, Holzer G, Isensee K et al (2019) Dealing with the effects of ocean acidification on coral reefs in the Indian Ocean and Asia. Reg Stud Mar Sci 28:100560
Leong C (2015) Persistently biased: the devil shift in water privatization in Jakarta. Rev Policy Res 32(5):600–621
Mishra BK, Emam AR, Masago Y, Kumar P, Regmi RK, Fukushi K (2018) Assessment of future flood inundations under climate and land use change scenarios in the Ciliwung River Basin, Jakarta. J Flood Risk Manage 11:S1105–S1115
Noorduyn J, Verstappen HTH (1972) Pūrnavarman's river-works near Tugu. Bijdragen tot de taal-, land-en volkenkunde 2/3de Afl: 298–307
Octavianti T, Charles K (2018) Disaster capitalism? Examining the politicisation of land subsidence crisis in pushing Jakarta's seawall megaproject. Water Altern 11(2)
Padawangi R (2012) Climate change and the north coast of Jakarta: environmental justice and the social construction of space in urban poor communities. In: Urban areas and global climate change. Emerald Group Publishing

Padawangi R, Douglass M (2015) Water, water everywhere: toward participatory solutions to chronic urban flooding in Jakarta. Pac Aff 88(3):517–550

Pawitan H, Baharsjah JS, Boer R (1999) ENSO impacts on Indonesia seasonal rainfall. In: Asian Pacific FRIEND and GAME joint workshop on ENSO, floods and droughts in the 1990s in Southeast Asia and the Pacific. Hanoi–Vietnam, pp 23–26

Podomoro A (2014) Pluit Cit. http://www.pluit-city.com/. Accessed 31 Jan 2014

Putri PW, Moulaert F (2017) Spatial practices and the institutionalization of water sanitation services in southern metropolises: the case of Jakarta and its Kampung Kojan. Int J Urban Reg Res 41(6):926–945

Rahardjo P (2017) Ecosystem approach studying development of North Coast Jakarta. IOP Conf Ser Earth Environ Sci 79(1):012009

Rukmana D (2015) The change and transformation of Indonesian spatial planning after Suharto's new order regime: the case of the Jakarta metropolitan area. Int Plan Stud 20(4):350–370

Siders AR (2019) Managed retreat in the United States. One Earth 1(2):216–225

Simanjuntak T (2006) Indonesia–Southeast Asia: climates, settlements, and cultures in late Pleistocene. Comptes Rendus Palevol 5(1–2):371–379

Supari RM, Sopaheluwakan A (2016) Mapping drought risk in Indonesia related to El-Niño hazard. In: AIP conference proceedings, vol. 1730, issue 1. AIP Publishing, p 070001

Surbakti IM, Idroes IC, Simarmata HA, Firman T (2010) Jakarta city report. In Workshops of climate change vulnerability assessment and urban development planning for asian coastal cities. Bangkok, Thailand

Tambunan MP (2017) The pattern of spatial flood disaster region in DKI Jakarta. IOP Conf Ser Earth Environ Sci 56(1):012014

Tapper N (1999) Atmospheric issues for fire management in eastern Indonesia and northern Australia. In: ACIAR proceedings. ACIAR, 1998, pp 21–30

Texier P (2008) Floods in Jakarta: when the extreme reveals daily structural constraints and mismanagement. Disaster Prev Manage Int J

Wade M (2020) Island City: urban development, planning, and real estate in Jakarta. University of California, Berkeley

Whittaker JM, Müller RD, Sdrolias M, Heine C (2007) Sunda-Java trench kinematics, slab window formation and overriding plate deformation since the Cretaceous. Earth Planet Sci Lett 255(3–4):445–457

Whitten AJ, Whitten T, Soeriaatmadja RS, Soeriaatmadja RE, Afiff SA (1996) Ecology of Java & Bali, vol 2. Oxford University Press, Oxford

Wijaya IMW, Soedjono ES (2018) Domestic wastewater in Indonesia: challenge in the future related to nitrogen content. Geomate J 15(47):32–41

# Chapter 4
# Istanbul, Turkey

**Abstract** This chapter presents the model of Istanbul as one that appears, with Sao Paolo and Seoul, to move beyond the crisis of severe inequality and informality to pose serious efforts to combat climate hazards. Istanbul's global leadership on the issue of water is shown, illustrating the efforts of the ruling party to develop some institutional approaches to flooding and drought management. The history of hydrologic infrastructure development offers a helpful perspective regarding the use of surface water to support the capital city's sprawling population. Stock is taken of the influences of climate teleconnections on the cyclical dry and warm periods experienced by Istanbul, and the impacts of climate change are discussed in relation to variability and intensity of extremes. As droughts intensify, the extension of water infrastructure to further surface water resources remains the natural, technocratic response tied to the continuation of past development policies. Yet, this chapter shows that newer approaches are being operationalized involving more advanced, participatory engagement with urban resilience. With political repression and authoritarian efforts to keep the ruling party at the helm of the nation's capital, however, such water policies face considerable obstacles amid the general pull of technocracy.

**Keywords** Teleconnections · Technocracy · Istanbul Water Consensus · Surface water · Tariffs

## 4.1 Introduction

In 2021, Turkey ratified the Paris Climate Accords, placing itself in alignment with others seeking to mitigate the impacts of climate change by proactively decreasing carbon emissions and increasing resilience. But Turkey's commitments to climate adaptation began more than a decade earlier, in 2009, when 40 mayors from around the world joined in signing the Istanbul Water Consensus (Ulibarrí 2011). Initiated at the Fourth World Water Forum hosted in Mexico City, the Istanbul Water Consensus expressed participants' "readiness to take leadership in advancing



H. Chang, A. R. Ross, *Climate Change, Urbanization, and Water Resources*, https://doi.org/10.1007/978-3-031-49631-8_4

integrated water management approaches to 'bridge divides for water' and strengthen the resilience of our cities and regions to cope with rising external pressures and contribute to our overall sustainable development" (Turkey 2009).

Together, the signatories affirmed their commitment to high water quality and sanitation, agreeing to keep water under public control while striving to increase water access and sanitation for poor slum dwellers as well as property holders. Noting similarities in water issues across regions, the consensus argues that "equitable, optimal and sustainable management of water resources and services demands an integrated approach, coordinated action and the sharing of responsibilities by the various tiers of government."

In addition, this coordinated action calls for "more water-sensitive cities," including "local sustainable water management" (Turkey 2009). While this consensus originated among mayors, it expanded to involve regional and national cooperation. By 2012, some 1000 mayors from 58 countries had signed up for the Istanbul consensus, illustrating Istanbul's role in promoting sustainable water management around the world.

One of the biggest megacities in the world, Istanbul's population lies well in excess of 15 million, with a growth rate of 250,000 per year. With a growth rate of 2.8% annually, Istanbul's population is growing at twice the rate of the country in general. In addition, the Syrian War caused an influx of refugees to Turkey of some 600,000 people per year (Edelman 2021).

This large population requires significant water resources, yet 98% of Istanbul's water, both industrial and residential, derives from surface sources (Ozturk and Altay 2015). A majority of this surface water comes from the Asian side of Istanbul, while a full 67% of the population lives on the European side. Today, water demand stands at over 175 liters per capita and will likely increase to 225 liters per capita by 2050 (Edelman 2021).

Yet, the growing dependence of Istanbul on water transfers from surface water basins outside of the city indicates the increasingly unsustainable practice through which the city has developed for hundreds of years. Beyond 2060, a study of water transfers conducted by Burak et al. shows that transfers and local surface water will not suffice to meet the needs of the growing population of Istanbul. According to that study, "water shortage cannot be resolved for Istanbul unless water transfer is accompanied by additional sustainable management policy measures… There is a need to adopt a more diversified water supply and demand management portfolio which have been implemented successfully in other megacities and have hedged supply risks" (Burak et al. 2022) (Fig. 4.1).

## 4.2 Climate and Physical Geography

Located on the Çatalka peninsula, where the Bosphorus Strait splits Europe and Asia, Istanbul is situated in a global position so crucial to commerce and transportation that under the Ottoman Empire it became known simply as "the Port." Given

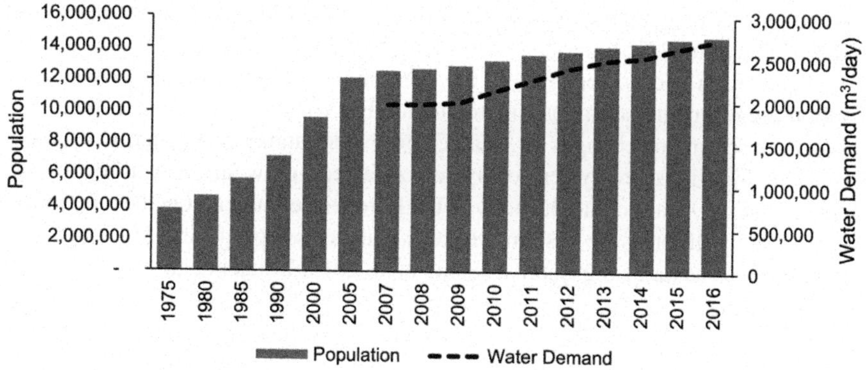


**Fig. 4.1** Population changes and annual water demand in Istanbul, 1975–2016. (Source: Cuceloglu et al. 2017)

such a position, one might imagine Istanbul to be one of the most water-rich places on Earth, but this is far from the truth. Despite its ideal strategic location, Istanbul has always faced significant pressures from water demand, and as the city continues to grow, climate change only enhances those pressures.

Situated on the 41st parallel, the Çatalka peninsula sits just north of the mid-latitude deserts that belt the globe and manifest regionally in the Arabian desert. For Turkey's southern region, this translates to arid and semiarid climates in the Anatolian steppe lands descending from the Taurus mountains toward the southern border with Syria and the western Mediterranean coast (Aksoy et al. 2020). Evaporation off the Black Sea feeds most of Turkey's precipitation over the belt of northern Anatolian mountains (Atalay 2006). Due to these mountain ranges, however, rain remains isolated in the oceanic climate of the Black Sea region, while the Southern and Western steppes remain relatively water stressed (Sariş et al. 2010).

In the northwestern corner of Anatolia sits the Çatalka peninsula, cut in half by the Bosphorus straits, formed through fault activity and erosion during the Holocene. Drained by rivers originating in the mountains to the southeast, the Çatalka holds few substantial freshwater resources in its eastern portion. To the west, in Turkey's Thracian region bordering Bulgaria on the Black Sea, a different picture emerges. Lake Durusu, also called Lake Terkos, lies 40 km northwest of the city and spans 25 km$^2$ in size, with further freshwater systems feeding verdant forestland in the area (Soylu 2009).

Yet, rain hardly falls evenly across time on the Çatalka peninsula straddling the Bosphorus straits between Europe and Asia. Unlike the northern Anatolian mountains, there are no consistent geomorphic features that force precipitation through an orographic lift, so Istanbul must rely on weather patterns from the far-away mid-Atlantic. More specifically, low-pressure systems caused by cold weather patterns near Iceland form a kind of gear-like mechanism with the high-pressure systems around the Azores Islands roughly 850 miles west from the coast of Portugal. The combination of two systems churns out westerly winds rich with water vapor, providing a critical source of rain for Europe and the Mediterranean (Küçük et al. 2009; Şahin et al. 2015).

The position and power of these low- and high-pressure systems vary periodically in a pattern almost impossible to predict. But when they do change, their variability causes major changes in weather over the Çatalka peninsula. On the one hand, the high-pressure system over the Azores can grow stronger, throwing fast winds from the Atlantic into the body of the European continent. On the other, when the Azores' high-pressure system weakens, winds swoop into the Mediterranean, bringing rains as far as the Çatalka (Şahin et al. 2015; Baltaci et al. 2020).

## 4.3 History of Water Management in Istanbul

Human civilization on the Çatalka has found that, when this North Atlantic Oscillation is positive and the Azores system is the strongest, droughts become a major obstacle for production, trade, and even subsistence. Established as Byzantine in the seventh-century BCE, the city grew to become Constantinople, the capital of the Eastern Roman Empire, after the schism with Rome in the fourth century. During this period, subsurface aquifers remained the most important source of drinking water, but Istanbul's distance from freshwater supplies led authorities to build aqueducts. Under Emperor Constantius II and Valens, Constantinople saw the construction of the Valens Aqueduct, which was expanded after the severe drought in 382 and ultimately extended some 430 km, drawing water from Pazarlı, 120 km to the west. Numerous covered and open cisterns also existed for storage, with a combined capacity of about one million $m^3$ (Crapper 2020).

Intriguingly, the climatological variability of the region may have brought about unrest, as tree-ring evidence suggests that the Huns and Avars may have migrated westward during the fifth century as a result of a particularly water-stressed period. Indeed, from 300 to 560 CE, Southern and Western Anatolia recorded about 12 major droughts—taken together, a serious dry period relative to the historic mean. While the aqueduct fell out of use as a result of conflicts involving the Avars and the growth of a relatively wet period from 560 to 730 CE, a dry phase set in again with the onslaught of another catastrophic drought in 758 CE, leading to the aqueduct's revival (Crow 2012; Haldon et al. 2014). By the turn of the second millennium, Constantinople's impressive hydraulic engineering was still sustaining what had become the largest population in Christian Europe.

Ottoman conquest in the fifteenth century, along with the mini-Ice Age, brought about a renewed appetite for grand construction and the development of new water lines over the course of successive sultans. The 1500s saw increasingly dry and cold weather, leading to the failure of crops in Anatolia, putting a strain on Constantinople's food and water supply. The Ottomans built supply lines consisting not merely of pipes or aqueducts but entire water systems involving lengthy lines supporting and supported by cisterns and wells, feeding networks of public fountains and famous Turkish baths (De Feo et al. 2013). Built around 1563, the most extensive of these, the Kırkcesme water supply system, is still in operation today and includes monuments such as the Maglova Aqueduct located between Lake Durusu and the city (Altinbilek 2006). Yet, in 1596, the terrible drought stoked the fires of the Celali Rebellion, which broke the empire's Golden Age and brought about an authoritarian backlash that changed Ottoman rule indelibly (White 2011).

## 4.4 Modern Water Management in Istanbul

Another extended dry period wracked the region in the nineteenth century, according to available tree-ring records, showing more frequent droughts. As a result, river systems converted to marshland, creating, along with deforestation, breeding grounds for malaria and other diseases, as water quality diminished resulting from increased concentrations of pollution. With hotter temperatures, supply chains based on pack animals began to break down. Prices for firewood, charcoal, and timber skyrocketed, while the Port leaned on the countryside harder to provide resources. Worsening sanitary conditions created by the increased prices of public baths contributed to the increase of disease, leading to a vicious cycle of supply chain deterioration. In 1868, Istanbul created its first municipal institution, the Şehremaneti, to cope with the sanitation problems brought about by the dry period (Sert 2022; Kentel 2018, 2021).

The following year, the city commissioned a French company called Dersaadet Water Company to provide water for the European part of Constantinople, bringing in the Uskudar-Kadikoy Water Company (also French) to serve the Asian portion in 1888 (Fleet 2018). The former company drew water from Lake Durusu and developed the first treatment facility in the city in 1926. In the 1930s, a new nationalist state took power and organized control over the city's water through a public company called the Istanbul Water Administration. It was believed that the public utility involving fees and tax revenues would better serve the people. In 1952, the Elmalı-2 Dam was built to the east of the city, and soon after, the government created the State Hydraulic Works (Altinbilek 2006).

From 1971 to 1974, Istanbul experienced another dry period, leading the State Hydraulic Works to consolidate renewed efforts into a Master Plan that would modernize infrastructure and expand capacity for service to the city's growing population. Within 10 years of the master plan, the state created the Istanbul Water and Sewerage Administration in order to address problems of water quality as well as quantity. Not only would the new administration rationalize and construct water supply and distribution, but it would also treat and collect wastewater, filter pollution, and restore waterways within the metropolis. Increasingly, water development concentrated on providing not only sufficient water but also better water as the country expanded its industrial production apace (Mutailifu 2019).

However, the onset of an extended dry decade from 1982 to 1993 threatened the water supply of the city even further. During this period of modernization, the water authority further constructed two large reservoirs on the Asian side of Istanbul—the Ömerli and Darlik lakes, both created by damming streams (Demirci and Butt 2001). Together with Lake Durukus and a reservoir created by the Alibeyköy Dam, these freshwater reservoirs continue to provide an important majority of the 97% of Istanbul's drinking water that comes from surface water sources.

Currently, the city's infrastructure accounts for 783 million $m^3$ of water, with the Istanbul Water and Sewerage Administration, alone, employing 6010 people and handling an operating budget of the public utility İstanbul Su ve Kanalizasyon İdaresi which was US$ 1.2 billion in 2005. Yet, despite the impressive infrastructural expansions of the twentieth century, by 2000, only 10% of Istanbul residents drank most of their water from the tap (Altinbilek 2006).

## 4.5 Climate Change

Water management in megacities is always a complex phenomenon, and Istanbul is no exception. With 11.5 million residents, a full one-sixth of Turkey's population resides in Istanbul, which houses 40% of the country's industry (Durukal et al. 2008). Since so much of Istanbul's water comes from surface sources such as reservoirs and rivers, preservation of those sources takes a top priority among the city's water managers. However, climate change presents a strong challenge to water resources management in terms of both quantity and quality, as the intensification of storms threatens to exacerbate flood damage, while the question of interdecadal variability that has impacted this region of Turkey over the centuries remains a key point of uncertainty.

## 4.6 Floods

The local Ayamama River is known to overflow its banks, leading to disastrous floods in 1995, 2009, 2011, and 2017. On September 9, 2009, for instance, a debilitating flood injured 50 people and killed 32, devastating houses and properties throughout the Marmara area and particularly the Istanbul suburbs. Some 35,000 people were impacted by the flooding (Kömüşcü and Çelik 2013).

Situated between the Kucuk Çekmece Lake and the Golden Horn, the Ayamama River became an essential landmark of Istanbul as the longest watershed in the region, running through the city all year around unlike most area streams. For more than a thousand years, the river provided ample drinking water through the aquifers that it drained beneath the burgeoning capital. And as the city grew, so did its water demands. With an area of 66.75 $km^2$, the Ayamama drainage continued to support even the growing industrial sector in Istanbul, leading to increased urbanization of the areas around the river (Janizadeh and Vafakhah 2021). With industrial pollution of the waterway and green spaces covered by impervious concrete and cement, the

Ayamama was no longer able to recharge Istanbul's vital aquifers with clean water, filling them instead with heavy metals that emptied into Kucukcekmece Lake, removing it as a basin for drinking water (Demiroglu 2019).

One study looking at the Olimpiyat climate station in Istanbul detected an increase in maximum rainfall amount over the course of the next 50 years, according to climate simulations ranging from no global decrease in fossil fuel emissions to some decrease in emissions. While the results vary depending on the simulation, the intensity of storms could increase more than 20%, the study claims, noting in particular the potential for flood risk in the Ayamama Watershed. This potential is made even more troubling by the land use and land cover change around the river during the 2010s' increased peak flows by nearly 20% (Nigussie and Altunkaynak 2019).

In addition, a study of data collected at the Florya climate station in Istanbul shows that climate change will drive an increase in the intensity of short-term rainstorms most of all. This study confirms the major problems of flash flooding in Istanbul in the foreseeable future, making further efforts to control rainwater runoff all the more critical. Particularly regarding polluted waterways such as the Ayamama, flash flooding can cause contamination in flood-prone areas, leading to health issues and disease. At the same time, the concentration of precipitation in short bursts makes groundwater recharge more difficult and does not slake the need for consistent freshwater supply (Güçlü et al. 2018).

## 4.7 Droughts

More than short-term rainfall intensity is needed to stave off the other side of climate change's impacts in Istanbul: drought. Due to higher temperatures brought on by climate change, the high-pressure system over the Azures will also grow more powerful. As a result, saturated water from the Atlantic will likely batter continental Europe with increasing floods more often, bypassing the Mediterranean and leaving Istanbul with more sustained drought periods. Combined with rising temperatures, this general change in the North Atlantic Oscillation will bring significant, long-term socioeconomic impacts to the region.

Since 1912, the temperature in Istanbul has risen by some 0.94 °C, a trend marked especially after the period of industrialization beginning in the 1940s (Burak et al. 2021). While precipitation has increased more generally, severe drought also struck in the mid-1980s and 1990 (Bakanoğulları and Yeşilköy 2014). The year 2006 saw the lowest precipitation rates in 50 years, indicating that the increased heat may add to variable patterns of dryness (Van Leeuwen and Sjerps 2015). Also, warming will provide favorable conditions for mosquitos, leading to more potential for disease. By changing the composition of the atmosphere, warming will also lead to a deterioration of air quality, which will put increasing pressure on the city's health, resulting in respiratory diseases, ailments such as asthma, allergies, and cancer, according to a 2020 study published in *Economic and Social Changes.*

Additional energy costs will be incurred to cool buildings, and crop development will likely be impacted (Aygün and Baycan 2020).

Across Turkey, soil moisture at plow depth has declined, particularly during the growing season when water is needed the most. Thus, irrigation has become increasingly essential for agricultural production. Yet, the increasing risk of extended drought makes the supply of surface water more tenuous—particularly in light of the water demands of hydroelectric electricity.

The worst recent drought in Istanbul began in 2006, tumbling from a meteorological phenomenon into an agricultural drought, ultimately culminating in a hydrological drought with massive socioeconomic impacts. The dire socioeconomic drought led to farming losses, diminished subsurface water resources, and faltering surface water reserves. Through the drought, Istanbul pulled back on water supply to residents, issuing a rationing policy and attempting to transfer water from nearby rivers (Kurnaz 2014).

But another meteorological drought hit in 2012, dropping precipitation levels to 63% of normal levels going into 2014. From 2012 to 2014, water transfers increased from 18.3% of the total water supply to 39.1% in order to compensate for the dry period (Burak et al. 2022). This drain on resources combined with the shrinking of Istanbul's dam-held reservoirs to half the levels of their previous year. Hydroelectric energy production faces challenges in light of such hazards, as 2014 saw water levels for Turkey's 88 energy-producing dams decrease to 45% (Kurnaz 2014).

According to climate models using CMIP5 simulations, the stretch of Turkey from Istanbul eastward along the Black Sea is set to experience the most summer drying of any other region in the Mediterranean, indicating that water conservation and supply are becoming increasingly urgent (Kelley et al. 2012). Thus, increased water transfers from neighboring basins could potentially spark conflict, given the increased need for local water use near those basins.

## 4.8 Water Management Strategies

Despite widespread recognition that climate change bears potentially serious implications for water resources management in Istanbul, major obstacles to change remain. The political leadership often swings from party to party within the city, bringing up new agendas and ways of approaching problems that take extended periods to institute. Also, these new agendas often come into tension with current laws, bringing about a lag in efforts to alter management policies. Lastly, urban development remains a crucial part of Istanbul's economic and demographic growth, but urbanization priorities often clash with those of climate adaptation measures such as expanded green space and water quality improvement.

According to a Soil and Water Assessment Tool (SWAT) model of Istanbul's water potential, while the Asian side of Istanbul holds more "blue water" resources from watersheds, containing 77% of the city's water supply, renewable "green water" resources remain prevalent in agricultural areas on the western, European side. These resources, which include soil moisture, have the potential for agricultural growth and economic revenue but may also be vulnerable to climate change (Cuceloglu et al. 2017).

Istanbul is also working to dig a canal that would bypass the Bosphorus Straits by connecting the Black Sea to the Aegean Sea via the Sea of Marmara. This enterprise would limit traffic in the Bosphorus, offering economic benefits as well as improvement to water quality in that important stretch of the region. However, additional environmental impacts on the ecology of Marmara may result (Gazioğlu et al. 2016).

Other important infrastructural projects are emerging in response to water quantity problems predicted in the event of a 2 °C increase in temperature, Istanbul's water supply could decrease by some 14%. This would increase the need for high-cost alternatives to the present surface water system, drawing more water from the Melen Basin 180 km to the east, which includes some 45% of the total water supply for Istanbul. Using a tunnel 6 m in diameter and 135 meters below sea level, some three million cubic meters of water passes through this route per day from the Melen Basin to Istanbul. Another 25% of Istanbul's water supply comes from the lakes Omerli, Terkos, and Buyukcekmece, yet because Omerli lies within the urban area, protecting water quality will become increasingly important (Fig. 4.2).

Luckily, wastewater management has greatly improved in Istanbul. With the population more than tripling from 1980 to 1995, untreated wastewater discharged into the Sea of Marmara, the Bosphorus, and the Golden Horn became a growing problem. Especially around the Golden Horn estuary, with its 700 industrial sites, static waters grew anaerobic, causing foul-smelling contamination along the beaches of the region. A massive $653 million project undertaken in 1994 and finished in 2002 dredged millions of $m^3$ of sludge, shut down industrial sites, and created recreational areas led the way for a broader regional approach to contamination and wastewater management (Altinbilek 2006). By 2004, the amount of treated wastewater in Istanbul had increased from 9% to 95% in just over 10 years. From extinction, wildlife has improved to 33 species of fish found in the area now, and coliform bacteria value has improved from 350,000 per 100 ml to only 1000 (Van Leeuwen and Sjerps 2016).

Such efforts will be increasingly important as floods and droughts threaten water quality under climate change. Separation between sewage and storm water is particularly useful for designating some wastewater for reuse, while ensuring proper treatment of sewage (Yuksel et al. 2004).

In addition, scholars argue for increased conservation of water in the fields. Turkey uses some 70% of its freshwater for agricultural irrigation, which will become more crucial as climate change prolongs drought. However, farmers supply much of this irrigation water to their fields through wasteful practices such as flooding. Also, irrigation water is not measured by farmers, leading to precarious situations of waste. Since rainwater is less available and irrigation more important, scientists argue that greater control over freshwater supply to farms will enable greater access to drinking water for growing megacities (Kurnaz 2014).

Moreover, conservation efforts can be increased through "water sensitivity." In the words of Cuceloglu et al. (2017), "In addition to increased water potential

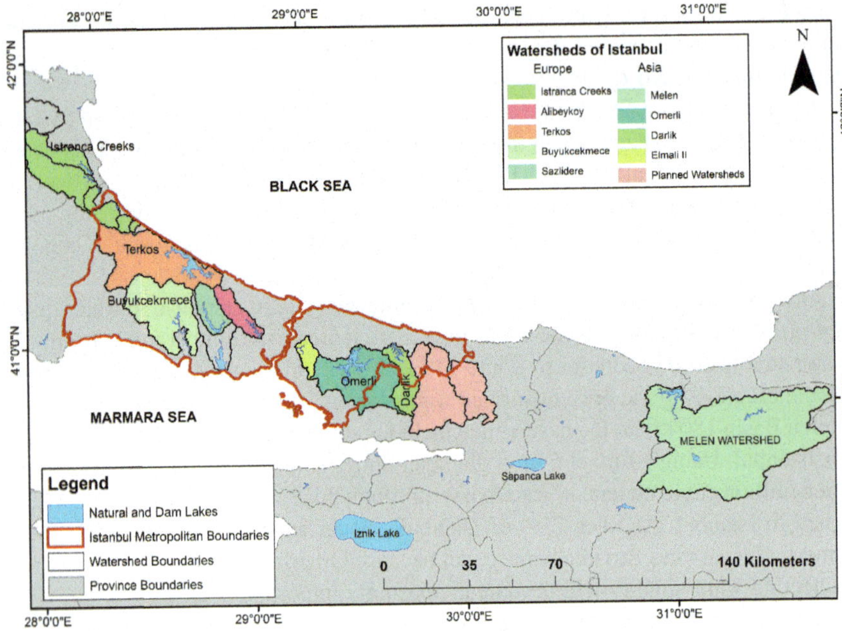


**Fig. 4.2** Watersheds that provide drinking water for Istanbul. (Source: Cuceloglu et al. 2017)

capacity, water reuse, decreasing water loss in the supply network, upgrading urban drainage system, rainwater harvesting, and the efficient use of available water will make better use of available water in İstanbul." Also, the constraints on freshwater access created by the needs of hydroelectric power can be lightened by switching to other forms of renewable energy, such as solar and wind. Other proposals for water conservation involve saving higher-quality water for residential uses, while retaining lower-quality reserves for industrial purposes, or recycling water for purposes other than drinking (Beler-Baykal and Oructut 2017).

While important, these large-scale engineering ventures address only part of the requirements of Istanbul to address its water needs. According to a study using the city Blueprint framework for assessing water capabilities, Istanbul lacks capacity in terms of energy and voluntary engagement in water strategies. Top-down approaches can aid in improving water issues, but a more mobilized public will be necessary to institute the kind of full-scale water policies made necessary by climate change (Van Leeuwen and Sjerps 2016).

## 4.9 Conclusions

It appears that the initiatives that began with the Istanbul consensus in 2011 have borne significant fruit, as the megacity experiments with a variety of different approaches to mitigating the causes and effects of climate change. In 2019, Istanbul's

Master Plan even set forward new targets in keeping with "Water Sensitive Cities," which have been pioneered in Melbourne and Singapore and would involve "a concurrent emphasis on supply and demand management, wastewater and stormwater management, institutional effectiveness and creating an enabling environment" (Burak et al. 2022). However, Istanbul's future is in question, as water politics plays a crucial role in competition between combative political parties.

The same year Water Sensitive Cities was introduced as a plank in the city's master plan, Istanbul voted in a new mayor, Ekrem Imamoğlu. From the opposition party, Imamoğlu represents a challenge to the ruling Justice and Development Party (AKP), which governed Istanbul for the previous 25 years. In the immediate aftermath of Imamoğlu's successful election, public agencies began to boycott bottled water from the Istanbul Metropolitan Municipality. As water costs doubled going into 2022, Imamoğlu attempted to raise water tariffs by 50% in an effort blocked by the AKP in the municipal council. Imamoğlu labeled the refusal "criminal" in just one of the verbal retaliations that brought about a two-year jail sentence for insulting politicians, passed down at the end of the year (Duvar English 2022). While Imamoğlu fights for his political career in the appeals courts, the future of utilities and water management remains opaque.

Istanbul shows that there is no "one-size-fits-all" solution; large projects can sometimes serve immediate needs, but sustainable solutions tend to require more active, local involvement involving deep, critical thinking about the role of water in the city. Yet, there is much left to be done. In July 2023, Istanbul's water consumption hit a record high, while its dams fell to just 40% total capacity—some dams had fallen below 5% at the time of writing (Daily Sabah 2023a, b). Istanbul faces the reality of becoming a "water poor" city in the next several years, indicating that adaptive measures to reduce water use among both residents and industrial stakeholders remain pressing.

## References


Aksoy E, Keskin S, Aktuz C, Bozdemir F, Muchoney D, Ozbek AK (2020) Mapping Anatolian Steppe region and ecosystem types by using earth observation and GIS PROCEEDING BOOK:58

Altinbilek D (2006) Water management in İstanbul. Water Resour Dev 22(2):241–253

Atalay I (2006) The effects of mountainous areas on biodiversity: a case study from the northern Anatolian Mountains and the Taurus Mountains. Grazer schriften der geographie und Raumforschung 41:17–26

Aygün A, Baycan T (2020) Risk assessment of urban sectors to climate change in İstanbul. ECONOMIC AND SOCIAL CHANGES-FACTS TRENDS FORECAST

Bakanoğulları F, Yeşilköy S (2014) Determination of meteorological and hydrological drought in Damlıca Creek watershed in Çatalca-İstanbul, Turkey. Türk Tarım ve Doğa Bilimleri Dergisi, 1(Özel Sayı-1), 1152–1157

Baltaci H, Alemdar CSO, Akkoyunlu BO (2020) Background atmospheric conditions of high PM10 concentrations in İstanbul, Turkey. Atmos Pollut Res 11(9):1524–1534

Beler-Baykal B, Oructut N (2017, June) Grey water as reliable/renewable source of water for cities of the future and an appraisal based on İstanbul, Turkey. IWA cities of the future conference: embrace the water, Gothenburg, Sweden
Burak S, Bilge AH, Ulker D (2021) Computation of monthly runoff coefficients for İstanbul. Therm Sci 25(2 part B):1561–1572
Burak S, Bilge AH, Ülker D (2022) Assessment and simulation of water transfer for the megacity İstanbul. Phys Geogr 43(6):784–808
Crapper M (2020) The valens aqueduct of constantinople: hydrology and hydraulics. Water Hist 12(4):427–448
Crow J (2012) Ruling the waters: managing the water supply of Constantinople, AD 330–1204. Water Hist 4:35–55
Cuceloglu G, Abbaspour KC, Ozturk I (2017) Assessing the water-resources potential of İstanbul by using a soil and water assessment tool (SWAT) hydrological model. Water 9(10):814
Daily Sabah (2023a) Istanbul water sources in danger as drought hits 3 crucial dams, July 2, 2023
Daily Sabah (2023b) Istanbul warns citizens against record-high water, consumption, July 17, 2023
De Feo G, Angelakis AN, Antoniou GP, El-Gohary F, Haut B, Passchier CW, Zheng XY (2013) Historical and technical notes on aqueducts from prehistoric to medieval times. Water 5(4):1996–2025
Demirci A, Butt A (2001, August) Globalization and water resources management: the changing value of water. In: University of Dundee International Specialty Conference, AWRA/IWLRI, August, pp 6–8
Demiroglu M (2019) Groundwater budget rationale, time, and regional flow: a case study in İstanbul, Turkey. Environ Earth Sci 78(24):682
Durukal E, Erdik M, Uçkan E (2008) Earthquake risk to industry in İstanbul and its management. Nat Hazards 44:199–212
Duvar English (2022) Istanbul mayor condemns AKP limit on water tariffs: 'What you are doing is a crime, February 18, 2022
Edelman DJ (2021) Managing the urban environment of İstanbul, Turkey. Curr Urban Stud 9(1):107–125
Fleet K (2018) The provision of water to İstanbul from Terkos: continuities and change from empire to republic. In: Middle eastern and north African societies in the interwar period. Brill, pp 212–238
Gazioğlu C, Akkaya MA, Baltaoğlu S, Burak S (2016) ICZM and the sea of Marmara: the İstanbul case
Güçlü YS, Şişman E, Yeleğen MÖ (2018) Climate change and frequency–intensity–duration (FID) curves for Florya station, İstanbul. J Flood Risk Manage 11:S403–S418
Haldon J, Roberts N, Izdebski A, Fleitmann D, McCormick M, Cassis M et al (2014) The climate and environment of byzantine Anatolia: integrating science, history, and archaeology. J Interdiscip Hist 45(2):113–161
Janizadeh S, Vafakhah M (2021) Flood hydrograph modeling using artificial neural network and adaptive neuro-fuzzy inference system based on rainfall components. Arab J Geosci 14:1–14
Kelley C, Ting M, Seager R, Kushnir Y (2012) The relative contributions of radiative forcing and internal climate variability to the late 20th century winter drying of the Mediterranean region. Clim Dyn 38:2001–2015. https://doi.org/10.1007/s00382-011-1221-z
Kentel KM (2018) Assembling 'Cosmopolitan' Pera: an infrastructural history of late ottoman İstanbul. Doctoral dissertation, University of Washington Libraries
Kentel KM (2021) Pera, Kasımpaşa, sewers, and maps: representing infrastructural entanglements in the nineteenth-century İstanbul. J Ottoman Turkish Stud Assoc 8(1):405–414
Kömüşcü AÜ, Çelik S (2013) Analysis of the Marmara flood in Turkey, 7–10 September 2009: an assessment from hydrometeorological perspective. Nat Hazards 66(2):781–808
Küçük M, Kahya E, Cengiz TM, Karaca M (2009) North Atlantic oscillation influences on Turkish lake levels. Hydrol Process Int J 23(6):893–906

Kurnaz L (2014) Draught in Turkey. IPC–MERCATOR Policy Brief. Istanbul Policy Center, Sabanci University, Istanbul

Mutailifu K (2019) Water and wastewater Management in the Mega City İstanbul: a general analysis from a supply-demand-reuse perspective

Nigussie TA, Altunkaynak A (2019) Impacts of climate change on the trends of extreme rainfall indices and values of maximum precipitation at Olimpiyat Station, İstanbul, Turkey. Theor Appl Climatol 135:1501–1515

Ozturk I, Altay D (2015, December) Water and wastewater management in İstanbul. In: Proceedings of the UNESCO HQ international conference on water, megacities and global change, Paris, France, pp 1–4

Şahin S, Türkeş M, Wang SH, Hannah D, Eastwood W (2015) Large scale moisture flux characteristics of the Mediterranean basin and their relationships with drier and wetter climate conditions. Clim Dyn 45:3381–3401

Sariş F, Hannah DM, Eastwood WJ (2010) Spatial variability of precipitation regimes over Turkey. Hydrol Sci J (Journal des Sciences Hydrologiques) 55(2):234–249

Sert E (2022) Metabolic flows of water in İstanbul in the nineteenth century: tap water, waste, and sanitation. J Urban Hist 00961442211073461

Soylu E (2009) Monogenean parasites on the gills of some fish species from lakes Sapanca and Durusu, Turkey. Ege Journal of Fisheries and Aquatic Sciences 26(4):247–251

Turkey (2009) İstanbul water consensus for local and regional authorities, World Trade Organization Committee on Trade and Environment

Ulibarrí N (2011) Bridging divides for water? Dialogue and access at the 5th world water forum. Water Altern 4(3):301

Van Leeuwen CJ, Sjerps RM (2015) The City blueprint of Amsterdam: an assessment of integrated water resources management in the capital of The Netherlands. Water Supply 15(2):404–410. https://doi.org/10.2166/ws.2014.127

van Leeuwen K, Sjerps R (2016) İstanbul: the challenges of integrated water resources management in Europa's megacity. Environ Dev Sustain 18:1–17

White S (2011) The climate of rebellion in the early modern ottoman empire. Cambridge University Press, Cambridge

Yuksel E, Eroglu V, Sarikaya HZ, Koyuncu I (2004) Current and future strategies for water and wastewater management of İstanbul City. Environ Manag 33:186–195

# Chapter 5
# Newcastle upon Tyne, United Kingdom

**Abstract** Newcastle presents among the most compelling models for confronting the impacts of climate change in the world today. Faced with the continuing problem of recurring floods throughout its history, Newcastle has adopted water management policies that integrate multiple ecosystem services in tandem with a policy approach that revitalized a deteriorating and depleted downtown area. This study shows how Newcastle changed its hydrologic plans from diverting water using walls and culverts to investing in blue-green infrastructure. Like Istanbul and Melbourne, Newcastle has helped lead cities' adaptive responses to climate hazards through a coherent and repeatable framework put forward through a decisive resolution. In particular, Newcastle's usage of nationally endowed funding affords space for interdisciplinary communities of researchers from public and private spheres to engage in open discourses on socio-ecological issues, including water resources management. These efforts have not only brought more life to the inner city but contributed to new ways of thinking about approaches to flood control, developing novel techniques for holistic and integrated water management focused on improving quality and quantity by joining urban and rural sustainable designs through runoff attenuation features and with sustainable urban drainage systems.

**Keywords** Sustainable urban drainage systems · Runoff attenuation features · Learning and action alliance · Blue-green infrastructure · Deindustrialization

## 5.1 Introduction

In February 2016, the Newcastle City Council passed a declaration alongside five partners—Northumbrian Water, the Environment Agency, Newcastle University, Arup, and Royal HaskoningDHV. Together, these groups committed themselves to "expanding the amount of Blue-Green Infrastructure in towns and cities across the UK," affirming that the economic threat of flooding merited "continued effort" to invest in the "physical and mental health and wellbeing (Newcastle City Council et al. 2016), biodiversity, carbon emissions, culture, quality of life and the



H. Chang, A. R. Ross, *Climate Change, Urbanization, and Water Resources*,
https://doi.org/10.1007/978-3-031-49631-8_5

economy" through infrastructure. Since investing in blue-green infrastructure would prove less costly than repairing flood damage, the signatories agreed on the necessity of not only "developing a supportive policy framework for new and retrofit projects" but "piloting new ways of working, and new funding models which help to realize the multiple benefits of blue-green infrastructure. Through "raising awareness and building capacity amongst communities," they argued that they could help with "providing local, regional and national leadership, encouraging and collaborating others" (Newcastle City Council et al. 2016).

Inspiring new projects and policies across the United Kingdom and making waves around the world, the Newcastle Declaration on Blue-Green Infrastructure worked so well that it was reaffirmed 3 years later, in March 2019, with an additional four signatories. Storage ponds, green roofs, green walls, and water channels would all play a significant role in flood control, along with a new water research facility housed in Science Central at Newcastle's Urban Sciences Building to test "smart" technologies for managing floods more effectively and with more precision.

The idea of blue-green infrastructure began as a useful effort to stave off the harmful impacts of floods on property as well as people's schedules and ways of life. By 2019, however, it had become part of a transformative vision of planning and design of urban spaces. With the University of Newcastle teaming up with other private sector partners to help produce a 24-acre development in the city center focused on applied and commercialized sciences called the Newcastle Helix, the city has become a veritable basin of attraction for laboratory sciences, business investment, and intellectual activity aspiring to increase resilience to climate hazards by generating transformative responses (Wright 2021).

## 5.2 Climate and Physical Geography

Newcastle sits on the Tyne River in the northeast of England near the border with Scotland and just upstream from the North Sea coast. Beginning around 310 million years ago, plant life in these coastal areas started to decay and form sediments deposited around the region in what became vast seams of coal, layered over with limestone, mudstones, cementstones, and sandstones during the hot Permian age around 60 million years later. As sea levels rose and mixed calcium and magnesium with the sands of the area, layers of Dolomite formed. The receding of the shallow "Zechstein Sea," which once spread across Europe, left that magnesian limestone over the coal (Smith 1989). However, with the change in climate from hot and dry to cold and wet, erosion from repeated, intense storms laid the seams bare over the course of eons, revealing the coal at the surface and creating the conditions for the eruption of economic power (Mawson Tucker 2009).

Two rivers—the Tyne and the Wear—cut through the coal measures of England's northeast, but the Tyne was afforded the advantage for transit by its more central location and length. Extending from high in the igneous Pennines, in the shadow of the Great Whin Sill formed through volcanic intrusion almost 300 million years ago as a result of tectonic tensions, the Tyne catchment drains nearly 3000 square kilometers of rich grasslands, fields, and forests (Johnson and Dunham 2001). Like its famous Scottish neighbors, the Tyne's 4000 km of riverways provide fruitful habitat for salmonids such as wild brown trout and grayling while linking the surrounding communities of Bellingham in the north fork, Alston in the south, Hexham near the confluence, and Newcastle closer to the mouth (Beechie et al. 2012).

Stretching eastward from Cumbria and Northumberland across northern England's north, the Tyne also provides a natural border that offered defensive purposes to the occupying Roman forces, as well as logistical obstructions. The Romans began to build bridges across the river at the beginning of the first millennium, with the Pons Aelius apparently intended as the end of Hadrian's Wall in modern-day Newcastle (Graafstal 2021). Among the most cherished artifacts of old Newcastle are pieces of Hadrian's Wall, a lengthy barrier across the Whin Sill meant to keep out Scottish raiding parties. Yet, Newcastle's significant strategic importance on the Tyne led to the need for enhanced fortifications.

Newcastle's climate is typically warm and temperate, with the hottest average temperature coming in July and rarely exceeding 29 °C. The least precipitation falls in March, followed by the outbreak of spring rains in April, which is the month that usually sees the most precipitation. Rainfall is fairly reliable throughout the year, however, as easterly winds cool adiabatically while rising up the Pennines, leading to saturation and precipitation throughout England's northeast (Fowler and Kilsby 2002). Because the Tyne basin begins in this range and descends through steep incisions of sandstone, limestone, and shale, the river system tends to respond rapidly to rainfall, flowing with high power through the system, especially during floods (Rumsby 1991).

The fairly frequent fluvial floods tend to increase with intensity during cooler, wetter years, decreasing with intensity during warmer, drier periods. Due to the inverse relationship between flooding and temperature, one might infer that warming trends patterned by anthropogenically induced climate change could decrease the total effect of flooding in the Tyne catchment. However, intense flooding in recent years, including 2017, would indicate that the temperature change may be counteracted by the increased evaporation rates leading to greater quantities of water vapor advecting from warming oceans. At the same time, 2022 saw the highest temperatures on record for England, with Newcastle lying in the upper reaches of the area impacted by temperatures 2 °C higher than the 1991–2020 average (Rumsby and Macklin 1994; Bertsch et al. 2022). Thus, Newcastle may experience hotter and drier summers without necessarily finding relief from cool-season flooding.

## 5.3 Water History

The fort built by the Romans to secure the Pons Aelius became the basis for an improved fortification constructed by the Saxons. The City of Newcastle likely attained its name as a result of the eleventh-century Castle Keep built out of wood on the same site and installed to curtail devastating incursions by the Danes, who occupied much of the territory to Newcastle's south. According to notes taken by John Leland while traveling to England in the mid-sixteenth century, Newcastle's medieval citizenry developed piped water systems through a process of charitable contributions and civic collaboration. Like others throughout Europe, the medieval town utilized water for various tasks, such as manufacturing, food processing, washing, drinking, cloth making, waste disposal, and milling grain, as well as the aforementioned defensive purposes. Yet, these often contradictory uses put city managers in a bind, leading to difficult decisions. For instance, during the late fourteenth century, the nearby City of York issued an ordinance to prevent butchers from disposing of carcasses in the river Ouse to conserve drinking water quality. In that instance, the butchers were left with the bill (Lee 2014).

In general, the earliest method of piped water appears to have stemmed from the Franciscan friary, who created conduits for water extraction. Scholar John S. Lee describes the conduits in his 2014 article on piped water access to medieval towns: "A conduit head was usually placed over springs with a cistern to collect water. The conduit was laid to carry the water—through pipes or an artificial channel—and a fountain-like structure erected (also known as a conduit) from where water was distributed to consumers" (Lee 2014). Medieval English water infrastructure tended to avoid open channels, preferring instead airtight pipes that, when filled with water, could transport its contents up hills without any pumping or feats of engineering, provided only that the conduit head lay lower than the source. In some cases, cities used horse-driven water wheels that drew water from wells with buckets that poured into a piping system.

Friars tended to share water conduits with burgesses in a kind of mutual aid arrangement, but both stakeholders jealously guarded their conduit houses from diversions beyond their settlement. In 1341, the Newcastle friars complained that townsfolk had finally revolted, breaking into their conduit house and diverting the water flow (Lee 2014). In contrast to the somewhat grudging agreement between the friary and the burgesses, commoners rarely gained access to the water franchise, thus making this early manifestation of direct action somewhat novel. Of course, this incident came at the beginning of the Hundred Years War and, perhaps, foreshadowed the massive insurgency of peasant unrest that would take London by storm 40 years later.

In the fifteenth century, burgesses in cities such as London, Southampton, and Cambridge increasingly agreed to extend piped water to the lower class city dwellers, probably due to these sorts of pressures from below, and pumps started coming into use later on in the century. Yet, municipal authorities still attempted to regulate use of the growing water supply, restricting quantities exploited by brewers and trying to protect local sources from contamination. By the mid-seventeenth century, Newcastle boasted five conduit heads around the city (Lee 2014).

By this point, England had left the Middle Ages behind, embarking on an early-modern period characterized by Tudor centralization and economic transformation. Under the Tudors, the English yeomanry gained landholdings and powers that brought an increasingly ambitious civic life to bear amid the Reformation of the

Church. In 1532, the General Act Concerning Commissions of Sewers passed parliament, codifying older practices of drainage and flood management in existence since at least the thirteenth century. Commissions of Sewers are now convened throughout the country, staffed by propertied residents with a given amount of money in goods. Thus, yeoman farmers and nobles joined together to address disputes over water, maintain water infrastructure, and report nuisance flooding (Morgan 2017).

While the Commissions of Sewers fell afoul of some local authorities, who saw them as bureaucratic impediments to municipal administration, they did provide a systematized way to address repeat flooding on a continuous basis, rather than setting up ad hoc commissions after each flood (Morgan 2017). The benefits of such a sustained effort would have been deeply felt in Newcastle, where the Tyne's semi-regular flooding presented a real problem. In 1339, a massive flood of the Tyne rushed the city, breaking through its city walls and killing a hundred residents (Hearnshaw 1924). This was an extreme example but shows the level of threat that flooding posed to the area.

In this sense, Commissions of Sewers provided an early model of modern state making, bringing the "middling sort" into administrative roles that issued from a central power controlled neither by the nobility, clergy, nor the crown, but by rational and codified rules. Thus, by integrating members of early Independent congregations, Presbyterians and even Quakers (provided they agreed to remove their hats, which was not always a given), water administration took an important early role in the development of a kind of sovereignty later leveraged against the crown by developing governance experiences for a politically active citizenry. In addition, Newcastle gained a fresh water supply from the springs at Gateshead, south of the Tyne and thus higher than much of the surrounding area. A second water source was added from Heworth Fell, via a trench and wooden pipes leading to two ponds, where lead pipes brought it into town. So, by the beginning of the seventeenth century, the city maintained both an active water management administration and new sources of freshwater, but the good times would not last long (Morgan 2017).

During the seventeenth century, however, efforts were undertaken to archive and document the customs and practices of the Commissions, leading to the disempowerment of commoners who lacked the skills to read, write, and research archival material (Skelton 2017). The centralization of water authorities regarding floods, sea defense, and drainage became, by the Industrial Revolution, a matter of expertise, as traditional customs and practices faded into knowledge sequestered in institutional archives (Crooks et al. 2002). Such a change in administration would be felt 100 years later, when a devastating flood washed away some 20 bridges on the Tyne, killing several residents and inundating the City of Newcastle.

It started on November 16, 1771, when a deluge hit the Pennines, swelling the Tyne to an enormous level perhaps not seen since. One chronicler named William Garret wrote of "the water in the Tyne rising six feet higher than a remarkable fresh in 1763." Garret continued, "The first dawn of day discovered a scene of horror and devastation, too dreadful for words to express, or humanity to behold, without shuddering: all the cellars, warehouses, shops, and lower apartments of the dwelling-houses … were totally under water" (Garret 1818). The flood swept away not only the bridges along the river but also countless livestock and animals, leaving the riparian floodplain scoured in its wake (Fig. 5.1).

Even as rebuilding commenced on the bridges over the Tyne, new floods hit in 1782, 1815, 1833, and again in 1835 (Rennison 2001). The 1839 flood broke through the wall again, rising up to 3 feet in some areas, while the one before it similarly brought flood levels to about 2 feet. Yet, the second half of the nineteenth century saw the highest peak of intense flooding in recent history. Specifically, intense flash floods in 1872, 1890, and 1913 caused significant turmoil. The first of the three events saw brutal hail crash through windows around the city, withering thunderstorms sending bolts of lightning that claimed three lives and buckets of water rising up to waist level. The 1890 flood logged repeated thunderstorms with five-day totals of 56–67 mm of flooding. Chronicles tell of "feet deep" water "rushing along like a river." Twenty-three years later, another incident produced 67 mm of rain in just 90 minutes at the Town Moor. In low-lying, poorer areas, almost every shop saw some flood damage, while carts stalled in waters up to their axles. The rushing water ripped up flagstones and sent them floating through the city, while the level of submergence of the market made the area totally inevitable except by small boat (Archer et al. 2016).

These incidents appear to have been flash floods whose onset came rapidly and surprised the population. They prompted efforts at flood mitigation, including flood walls and culverts to try to prevent flood waters from moving unimpeded through

**Fig. 5.1** An image of the ruined Tyne Bridge after the Great Flood, 1771, from "John Brand's History of Newcastle"

the city. However, the relatively rudimentary structures failed even in the twentieth century; first in 1900 after a 16-hour rainstorm brought nearly 80 mm of rain, and again 24 years later when 31 hours of rain and thunder spread out across the region, dropping 105 mm in Cockle Park alone, leading to surcharged culverts. Despite their massive impact, these incidents were largely forgotten by 1968, when a profoundly damaging storm dumped 72 mm of rainfall in just 4 h, leading again to the failure of culverts and the collapse of an upstream flood wall (Archer et al. 2016).

## 5.4 Flooding

It would seem that flood controls succeeded fairly well after 1968 until 2008, when a succession of three important floods hit, followed by two more in the ensuing two-year intervals. The year 2008 saw not only failed culverts but also manholes backing up into the streets, while 48 hours of solid rain brought 120 mm of flooding, leading to fluvial overflow of the Tyne river that impacted a thousand properties (Archer and Fowler 2018). In 2010, the Tyne again overflowed its banks at Morpeth, and in 2012, another powerful system caused widespread flooding. Indeed, the 2012 incident was so impactful that many saw it as totally unprecedented in the history of Newcastle. However, in the 200-year record, it likely ranked between the fourth and seventh largest flood, having likely been exceeded in 1941, 1913, and 1872, along with 1968, which ranks first overall. At the same time, “the sequence of recent floods suggests either a recent upward trend in natural variability or the impact of increasing temperatures on flood risk” (Archer et al. 2016).

One contributor to flood risk is land use change. Over the past century, deforestation has caused a decrease in rainwater infiltration, leading to more runoff. Agricultural production forced the transformation of geological contributors to river ways by altering slopes of hills and changing storage patterns, which led to the modification of connectivity, flow velocity, and timing. Also, the introduction of dense soils for agriculture increased lateral flows while decreasing vertical infiltration (Rogger et al. 2017).

Such findings indicate that climate change may not be responsible for the intense flooding seen from 2008 to 2012, and in fact, the late twentieth century saw a relatively low amount of flash flooding compared to the later part of the nineteenth century and early twentieth century. This indicates that Newcastle will likely have to expect greater flooding over the next 50 years than experienced prior to 2008, and that while the increased flooding may be influenced by hotter temperatures, natural variability also appears responsible for prior floods that rose to the level or exceeded the flood of 2012 (Pregnolato et al. 2017). Despite uncertainty surrounding the impacts of climate change in northern England, the expert consensus is that flood controls will be increasingly important in the coming decades.

## 5.5 Adaptive Strategies

It is, then, with good reason that Newcastle University partnered with the Environmental Agency after the 2008 flood to plan out small-scale management schemes. Among the plans set into place, the implementation of a new pond network offered the interesting appeal of providing some 20,000 cubic meters of storage, thus transforming the nuisance of floodwaters into a usable resource. These Runoff Attenuation Features would reduce flooding by up to 30 percent. The pilot for the program took place in Belford, about 40 miles north of Newcastle, which experiences negative impacts from flood events that hit the area but lacks the high-value housing and space for expensive or traditional flood prevention (Quinn et al. 2013a, b).

Runoff attenuation features (RAFs) are ideal for areas such as this for a variety of reasons. Their small size fit nicely in the hillscape of the green countryside without encumbering agriculture. Indeed, they can even aid farmers, both by providing convenient water sources and by assisting with drainage from cropland that might otherwise become supersaturated. These utilitarian features also function as a compliment to other flood mitigation approaches such as flood walls, rather than creating either/or situations. In addition, features can range in kind, from walling off a small area for ponding in the event of overland flow to digging ditches that lead to backwater fill, as well as stream diversions into temporary storage. Even the embedding of large wooden debris into streams can help increase hydraulic roughness while aiding in the creation of salmonid spawning habitat (Quinn et al. 2022).

Like Sustainable Urban Drainage Systems (SUDS), RAFs offer ecosystem services rather than merely playing roles restricted to the mitigation of flood damage and increasing resilience through the built environment. In rural areas, RAFs can help with the diffusion of agricultural pollution along with the more direct purpose of increasing the sustainability of farming by decreasing the harm of field flooding. While Newcastle remains at the forefront of RAF development, the expansion of these mechanisms to greater roles in the realm of water quality is still in the planning stage. However, the potential of RAFs to not only improve resilience but also introduce transformative processes to agriculture and rural areas, more broadly, remains compelling (Quinn et al. 2013a, b).

## 5.6 Deindustrialization and Renewal

As a result of deindustrialization beginning in the 1960s, Newcastle experienced a significant decline in population and urban development (Fig. 5.2). A city built on coal mining, parts of Newcastle's population fell on hard times when coal production declined in the United Kingdom. As the coal industry closed mines, a housing initiative called the New Towns project grew to popularity, establishing new, planned communities in suburban areas largely in the south. People flocked away from the

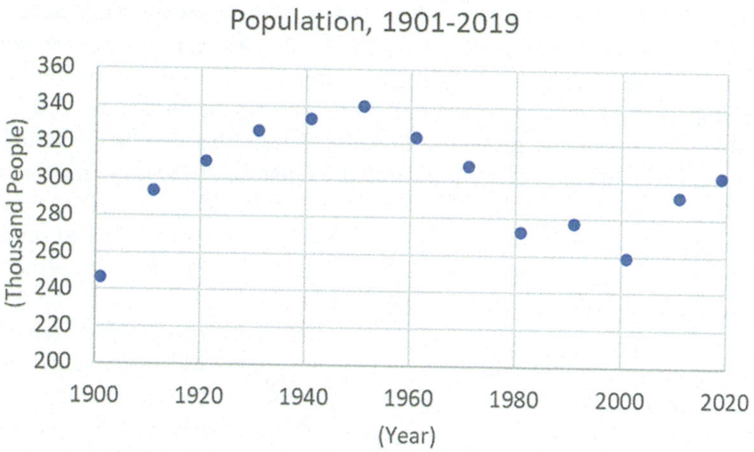


**Fig. 5.2** Population change in Newcastle, 1901–2019

dirty and failing coal industry toward new economic promises among the idyllic new cities (Alexander 2009). Of course, the housing projects were often less than idyllic, and the industrial north suffered a feedback loop of economic decline, in which people left the cities to avoid the problems of inner-city life, thus compounding the problems through capital flight. In the 1980s, coal miners went on strike to keep mines open and stop the collapse of the industry, but the government fought back and defeated them. The English steel industry fell apart soon after, continuing the urban decline (VAll 2007). As a result of the period from 1960 to 2000, Newcastle metropolitan area lost about a quarter of its population, descending from nearly one million residents to just under 750,000.

In part because of this experience, Newcastle University, which remained a strong public research university through the turmoil, worked to promote a visionary model of sustainable urban renewal based on economic development apart from the fossil fuel industry. In 2005, Prime Minister Gordon Brown declared Newcastle a Science City, conferring funding and networking opportunities on city administrators with the support of the University (Gertner and Bossink 2015). In 2005, the famous Newcastle Brown Ale company moved its brewery to Yorkshire, demolishing its works 3 years later. The city moved to buy the brownfield site with the University and develop "Science Central," the first of several buildings that would foreground urban resilience through partnerships with private partners dedicated to scientific research with a focus on climate change and other hazards.

In 2012, the City Council joined forces with the Gateshead Council and other partners to work out a new coordinated plan to implement SUDS throughout the urban area. Called the Newcastle-Gateshead Surface Water Management Plan, this effort promised to ensure that drainage systems can keep pace with development while preventing development from impugning drainage (O'Donnell et al. 2017). Built during the 1970s, the Howdon Sewage Treatment Works in North Tynside became the largest sewage treatment plant on the East Coast of England. Designed to remove overburden from River Tyne, the Howdon Sewage Treatment Works reached its capacity during the 2000s, and by 2012, it had come to lack the requisite headroom to meet the needs of regional development. For this reason, the Surface Water Management Plan posed a solution by relieving the burden of surface water from floods, delivering it instead to brownfield sites that can be converted from former industrial use into green infrastructure (Alstead 2011, Lawson et al. 2015) (Fig. 5.3).

Mandating developers to "remove as much surface water from joined sewer systems as possible, managing the water on site," the Surface Water Management Plan delegates responsibility for water management from the city, itself, to the development level. At the same time, planners intend to maintain a viable "management train" that echoes natural processes by utilizing SUDS in a coordinated fashion, from prevention at the site of development to control at the source, the local area, and the region (Alstead 2011). By combining these levels in a holistic strategy intended to reduce the volume and rate of runoff, while increasing water quality and adding to wildlife habitat in the urban area, the city becomes more attractive to humans and nonhumans, and more space is afforded in water infrastructure for flood resilience and residential growth, as well as agricultural sustainability in the city's outskirts.

**Fig. 5.3** An example of the Sustainable Urban Drainage System in the United Kingdom. (Photo by author)

In 2014, Newcastle developed the Learning and Action Alliance (LAA) framework, which would draw together different stakeholders hoping to conceive of flood management solutions based on governance that incorporates collaboration through system thinking. Newcastle was to become a "city that follows the principles of a Blue-Green City by maximizing opportunities to achieve multiple benefits of Blue-Green approaches to surface water management" (UFRRP 2019). The LAA helped to spread best practices throughout different social sectors while bringing people together for the assessment of shared goals. Participants cultivated atmospheres based on positivity and community, enabling greater civic cohesion and participation. At the same time, some participants came to worry that they were becoming an echo chamber, hoping to spread their thinking further in the future.

The efforts put forward by the city administrators and partners from 2008 to 2014—including Runoff Attenuation Features, Sustainable Urban Drainage Systems, and the Learning and Action Alliance framework—built on Science Central and set the path forward to progressive urban planning, which contributed to a larger process culminating in the Helix project produced in 2018. Reclaiming whole neglected city blocks in downtown Newcastle, the university launched an initiative to produce a site of scientific discovery with both private and public sponsors, building an infrastructure of laboratories, offices, lecture halls, eateries, and entertainment venues to draw residents into a vibrant downtown oriented toward generative synergies that refocus commerce and politics around a kind of green bourse—a marketplace of fresh, integrative ideas and practices that assemble together more participatory systems grounded in good will and community (Vallance et al. 2020).

## 5.7 Conclusions

While the impacts of climate change are complex as they pertain to Northeast England, indicating a mixture of floods and intensified droughts, Newcastle has led the way for future models of sustainable water management. Through a careful a long-term process of methodical, institutional development, Newcastle innovated both ideas for flood risk management and frameworks to get them done. Through this effort, Newcastle built on its established university structures, increasing its international prestige while drawing more people to a city that had lost a significant chunk of its population. In so doing, it exemplified value-based management.

In particular, sustainable urban drainage systems and runoff attenuation features play an important role in Newcastle's approach to curtailing flood issues. Rather than restricting the scope of water management to larger civil engineering projects, these alternatives can divert water into aquifers and away from the city center while increasing and using green spaces. Thus, instead of creating hazards for riparian biodiversity, including problems with water quality or quantity, the widespread implementation of decentralized and smaller-scale projects can increase community participation, enable groundwater recharge, and thus address different aspects of water management efficiently.

Indeed, these efforts to mitigate the effects of climate change have played an active role in making Newcastle the advanced and dynamic model for urban rehabilitation that it is today. The Newcastle Declaration remains, 8 years later, one of the most important international documents of modern water management, highlighting human–water dynamics in the urban context. With the Istanbul Consensus, it points the way to a collaborative path for addressing the complex problems that will characterize the future of urban sustainability and add a participatory element through which ends and means can be allayed.

## References

Alexander A (2009) Britain's New Towns: Garden cities to sustainable communities. Routledge, Abingdon

Alstead D (2011) Newcastle gateshead surface water management plan. AECOM, Newcastle

Archer DR, Fowler HJ (2018) Characterising flash flood response to intense rainfall and impacts using historical information and gauged data in Britain. J Flood Risk Manage 11:S121–S133

Archer DR, Parkin G, Fowler HJ (2016) Assessing long term flash flooding frequency using historical information. Hydrol Res 48(1):1–16

Beechie T, Pess G, Morley S, Butler L, Downs P, Maltby A et al (2012) Watershed assessments and identification of restoration needs. In: Stream and watershed restoration: a guide to restoring riverine processes and habitats. Wiley, Oxford, pp 50–113

Bertsch R, Glenis V, Kilsby C (2022) Building level flood exposure analysis using a hydrodynamic model. Environ Model Softw 156:105490

Crooks S, Schutten J, Sheern GD, Pye K, Davy AJ (2002) Drainage and elevation as factors in the restoration of salt marsh in Britain. Restor Ecol 10(3):591–602

Fowler HJ, Kilsby CG (2002) Precipitation and the North Atlantic Oscillation: a study of climatic variability in northern England. Int J Climatol 22(7):843–866

Garret W (1818) An Account of the Great Floods in the Rivers Tyne, Tees, Wear, Eden, Etc. in 1771 and 1815 With an Account of the Irruption of Solway Moss [by J. Walker].[Compiled by WG, Ie William Garret.]. Emerson Charnley

Gertner D, Bossink BA (2015) The evolution of science concentrations: the case of Newcastle Science City. Sci Public Policy 42(1):121–138

Graafstal EP (2021) The original plan for Hadrian's Wall: a new purpose for Pons Aelius? Archaeol J 178(1):107–145

Hearnshaw FJC (1924) Newcastle-upon-Tyne. Sheldon Press, London

Johnson GAL, Dunham KC (2001) Emplacement of the Great Whin Dolerite Complex and the Little Whin Sill in relation to the structure of northern England. Proc Yorks Geol Soc 53(3):177–186

Lawson E, Thorne C, Wright N, Fenner R, Arthur S, Lamond J et al (2015, November) Evaluating the multiple benefits of a Blue-Green Vision for urban surface water management. In: UDG Autumn Conference and Exhibition 2015, Leeds

Lee JS (2014) Piped water supplies managed by civic bodies in medieval English towns. Urban Hist 41(3):369–393

Mawson M, Tucker M (2009) High-frequency cyclicity (Milankovitch and millennial-scale) in slope-apron carbonates: Zechstein (Upper Permian), North-east England. Sedimentology 56(6):1905–1936

Morgan JE (2017) The micro-politics of water management in early modern England: regulation and representation in Commissions of Sewers. Environ Hist 23(3):409–430

Newcastle City Council, Northumbrian Water, the Environment Agency, Newcastle University, Arup, and Royal HaskoningDHV (2016) Newcastle Declaration on Blue and Green Infrastructure

O'Donnell EC, Lamond JE, Thorne CR (2017) Recognising barriers to implementation of Blue-Green Infrastructure: a Newcastle case study. Urban Water J 14(9):964–971

Pregnolato M, Ford A, Glenis V, Wilkinson S, Dawson R (2017) Impact of climate change on disruption to urban transport networks from pluvial flooding. J Infrastruct Syst 23(4):04017015

Quinn P, O'Donnell G, Nicholson A, Wilkinson M, Owen G, Jonczyk J, Barber N, Hardwick M, Davies G (2013a) Potential use of runoff attenuation features in small rural catchments for flood mitigation, NFM RAF. https://research.ncl.ac.uk/proactive/belford/newcastlenfmrafreport/reportpdf/June%20NFM%20RAF%20Report.pdf

Quinn P, O'Donnell G, Nicholson A, Wilkinson M, Owen G, Jonczyk J, ..., Davies G (2013b) Potential use of runoff attenuation features in small rural catchments for flood mitigation. Newcastle University (ed) Newcastle upon Tyne, Newcastle

Quinn PF, Hewett CJ, Wilkinson ME, Adams R (2022) The role of Runoff Attenuation Features (RAFs) in natural flood management. Water 14(23):3807

Rennison RW (2001) The Great Inundation of 1771 and the Rebuilding of the North-East's Bridges. Archaeologia Aeliana 29:269–292

Rogger M, Agnoletti M, Alaoui A, Bathurst JC, Bodner G, Borga M et al (2017) Land use change impacts on floods at the catchment scale: Challenges and opportunities for future research. Water Resour Res 53(7):5209–5219

Rumsby BT (1991) Flood frequency and magnitude estimates based on valley flood morphology and floodplain sedimentary sequences: the Tyne Basin, NE England (Doctoral dissertation, Newcastle University)

Rumsby BT, Macklin MG (1994) Channel and floodplain response to recent abrupt climate change: the Tyne basin, northern England. Earth Surf Process Landf 19(6):499–515

Skelton L (2017) Tyne after Tyne: an environmental history of a river's battle for protection 1529–2015. The White Horse Press, Cambridgeshire

Smith DB (1989) The late Permian palaeogeography of north-east England. Proc Yorks Geol Soc 47(4):285–312

Urban Flood Resilience Research Project (2019) Achieving Urban Flood Resilience in an Uncertain Future. University of Nottingham, Nottingham

Vall N (2007) Cities in decline?: a comparative history of Malmö and Newcastle after 1945. Malmö University, Malmö

Vallance P, Tewdwr-Jones M, Kempton L (2020) Building collaborative platforms for urban innovation: Newcastle City Futures as a quadruple helix intermediary. Eur Urban Reg Stud 27(4):325–341

Wright N (2021) The Newcastle Helix. In: The responsive university and the crisis in South Africa. Brill, pp 145–170

# Chapter 6
# Barcelona, Spain

**Abstract** Spain has a rich and textured history whose changes have shaped different water policies over the centuries—not always for the best. Situated in close proximity to freshwater resources, Barcelona's water infrastructure has evolved over diverse eras of technological capability and political control. This study discusses the climatological variability between floods and droughts, and the related generation of hydrologic adaptations to growing population and worsening climate hazards. It examines the water policies of the Francoist regime, including extensive water transfers for vast irrigation projects, along with the spread of water-vulnerable shantytowns, and also examines the way water issues fostered the social movements that ultimately brought down the Franco regime. In addition to municipalization, Barcelonans contribute to the participatory creation of blue-green infrastructure such as sustainable drainage systems throughout the city. While these projects help increase shade cover, improve soil condition, add wildlife, and control pests, they also bring up difficult issues of spatial inequality of income distribution and green spaces. The concern over what activists call "green gentrification" feeds into suspicions that reforms under the aegis of climate adaptation will only increasingly entrench the spatial inequalities of rich and poor in the city.

**Keywords** Municipalization · Climate extremes · Blue-green infrastructure · Green gentrification

## 6.1 Introduction

In a recent study on the impacts of floods and droughts covering 44 sites around the world, Barcelona emerged as one of only two sites that experienced successes. One of the scientists involved in the study highlighted two major reasons for Barcelona's success: "The improvement of risk management governance—more integration in emergency management and early warning systems—and the implementation of a series of structural measures that required high investment (the storm water reservoirs in Barcelona or the construction of dykes in Central Europe)" (Kreibich et al. 2022).



H. Chang, A. R. Ross, *Climate Change, Urbanization, and Water Resources*,
https://doi.org/10.1007/978-3-031-49631-8_6

The policies to improve water management in Barcelona did not occur overnight. An ongoing struggle involving memory and development, the deeply politicized process through which Barcelona's urban planners produced significant changes to adapt to pluvial flooding has historic roots in the city's origins and the history of the Spanish Republic (Popartan and Ungureanu 2022). Vigorous, ongoing debates over the city water's municipalization evoke discourses from the past that have, at times, been accompanied by violence. Yet, today they represent a need to reconcile adaptive development with the fissures of history in ways that residents, local administrators, and national authorities find acceptable.

Like the Istanbul Consensus, the Newcastle Declaration, and Melbourne's water-sensitive cities, the Barcelona model of adaptive management has taken an important place in the global popularization of water resources management practices that utilize modern technology and traditional knowledge, address local needs with an eye to larger-scale strategy, and employ self-managing communities to collaborate autonomously and effectively. As one of the most successful areas in Europe at ameliorating both floods and droughts, its programs and policies can be understood as vital, systemic responses whose novelty deserves further study and appreciation.

## 6.2 Climate and Physical Geography

A city of around 4.8 million people, Barcelona lies in northeastern Spain between the shores of the Mediterranean Sea and the mountains of the Catalonian Coastal Range. On both sides of Barcelona flow the Llobregat and Besòs Rivers, which drain the fertile coastal plains. Archeological artifacts of ancient coins suggest that, by the third-century BCE, pre-Roman inhabitants of the area called it "Bàrkeno," meaning "place of the terraces (or plains)" (Tinoco Mosteiro 2022). Here, on a hill, the Roman Empire established a military fortification, ultimately developing the larger municipal infrastructure of "Barcino" complete with a forum to act as the political center of the city. In the first-century ACE, the Roman authorities created aqueducts that would supply drinking water and irrigation to the city and its surroundings (Vázquez 2023; Orengo and Miró i Alaix 2013).

The city lies just 90 miles south of the peaks of the Pyrenees, the 310-mile band of mountains that separates modern-day France from Spain. While prior research suggests that the Pyrenees emerged through the collision of Iberia with continental Europe, a recent study indicates that more complex fractures and fissures set the stage for their rise. Namely, strike-slip faulting running east-west through the continental and Iberian produced a "pull-apart basin" called the Tardets-Sorholus trough, followed by activity from the transversal Barlanès and Saison listric, or curved, faults from southwest to northeast, which caused crustal thinning and subsequent exhuming of the mantle. By the late Cretaceous, the east-west strike-slip fault caused a north-south distention, producing the Mauléon Basin before its oblique contraction pushed up the Pyrenees range (Canérot 2017; Lagabrielle et al. 2010).

Jutting 150 miles south from the eastern edge of the Pyrenees, the Catalonian Coastal Range marks a continuation of the Iberian Massif, where ancient varisco rocks tell the tale of how the primordial Hunic superterrane smashed into Laurasia during the mountain-forming process. As mammals and birds evolved during the Neogene period, from these peaks that run parallel to the coast, rising more than a thousand miles high in the case of its Prelitoral alignment, tectonic pits of Campo de Tarragona, El Valles, and El Penedes filled with detritus, and the Valles-Penedes basin became inhabited by prehistoric apes and other species, giving us the "oldest Pliopithecus record in the world" (Instituto Geografico Nacional 2015; Alba et al. 2012).

Rising from glaciers and groundwater where the east Pyrenees hits the Coastal Range, the river Tor descends southward before cutting east into the Mediterranean for a length of about 20 miles, while the Besòs River flows from a confluence of tributaries in the Prelitoral range south into the sea. The Llobregat makes up the last of the three great Catalunyan rivers, extending from tributaries deep in the Pyrenees and the Coastal Range southeastward toward shores just north of what is now Barcelona city. These waterways gave life to migratory species from the beach to marshlands and coastal lagoons into pine forests, consisting of great cormorant and the endangered fartet fish.

These species thrive in the typical coastal Mediterranean environment, experiencing fluctuation between hot summers and cooler winters. Yet, at 41 degrees north, its latitude renders its climate more temperate than most of the rest of Spain. Throughout its history, though, the area has faced the problem of flash flooding. Especially during the summer, flooding hits the Pyrenees mountains to the north or the littoral mountains to the west, with instantaneous precipitation registering 180 mm/h—80 mm above the average. Impacts of flash flooding can include fluvial flooding of the Besòs and Llobregat Rivers, a problem that can occur even during dry periods to destructive effect. As recently as May 30, 2023, a flash flood hit Barcelona during the driest summer since the start of record keeping in 1961. Such a phenomenon reflects studies indicating that extremes—floods and droughts—may increasingly overlap one another as climate change persists.

## 6.3 Historical Development

As the city grew through conquests by the Visigoths in the fifth century, the Arabs in the eighth century, and the Carolingians in the ninth century, the aqueducts fell out of use but the site of the forum remained its center. It was not until the eleventh century that Barcino's water infrastructure expanded to include the Rec Comtal in the following century (Enrich Pericás 2022). During this period, however, groundwater resources from aquifers fed by the river deltas, as well as the rivers themselves, remained largely sufficient for the rate of urban growth and agricultural demands (Salgot and Angelakis 2019).

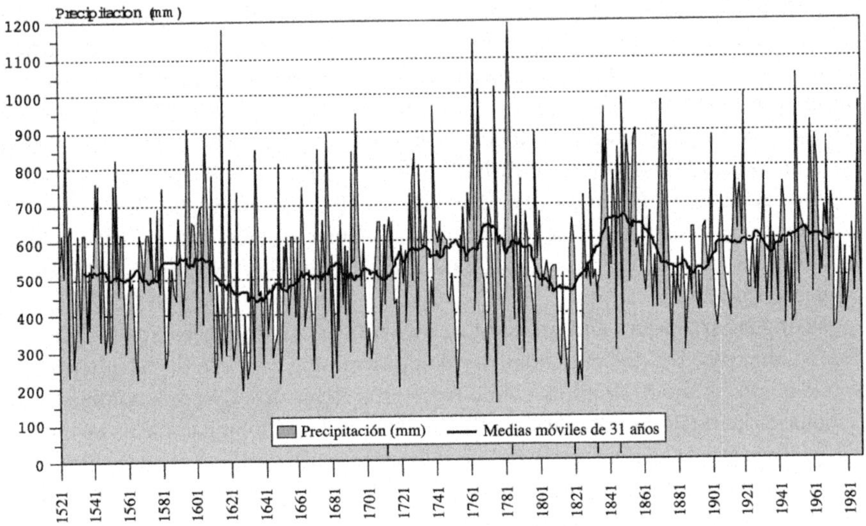


FIGURA 9. *Serie de Barcelona corregida (1521-1989).*

**Fig. 6.1** Precipitation variability in Barcelona, 1521–1989. (Source: Vallvé et al. 1998)

Municipal water systems were developed during the Middle Ages in the mid-fourteenth century, piping water flowing the Coastal Range from seven filtration galleries into a large tank called the Caseta de Jesus beyond the city walls. From the tank, the water flowed through pipes into drinking fountains distributed throughout Barcelona. However, pipe breaks, lime deposits, and problems with filtration systems led to an expensive maintenance bill, and climate variability led to water insecurity during dry periods (Guardia et al. 2014).

For the remainder of the medieval period, through the Renaissance and into the industrial era, Barcelona's water infrastructure remained fairly constant but also insufficient. Based on climatological proxy data provided by the written record of cathedrals, whose carefully documented *rogativas* marked prayer ceremonial prayers for the coming of new rains, the early seventeenth century experienced relatively more intense droughts in areas of Spain, followed by more rains in the late eighteenth century and another dry period in the early nineteenth century. Interestingly, the number of rain prayers correlated with the variation in the North Atlantic Oscillation (Vallvé et al. 1998) (Fig. 6.1).

## 6.4 Water and Development

Despite dry periods, during the eighteenth century, Catalonia's rich waterways and ample supplies of wool, along with the growing population of the urban center, made Barcelona an ideal Mediterranean site for the development of textiles (Vicente

2000). During the eighteenth century, the "crystalline water" provided the necessary resources for bleaching meadows next to mills in order to provide areas for the treatment of fabrics. As the cotton industry grew during the nineteenth century, Barcelona's mills used steam engines and, subsequently, its rivers, to "exploit hydraulic resources more thoroughly than any other place in Europe" (Thompson 2002).

Even in the late 1800s, as much as two-thirds of Barcelona's water was still derived from western springs, while the rest came from ephemeral streams and groundwater. Yet the city was growing, and new sources became necessary (Gorostiza et al. 2013). In 1879, a new water law ordained brought forward the use of steam engines to pump groundwater, mandating some 130 liters of water per day for urban use—somewhat more if the use of groundwater was involved (Masjuan et al. 2008). In 1882, the consortium, Sociedad General de Aguas de Barcelona, was created and soon gained the rights to begin pumping in the Llobregat basin (Matés 2019).

While by today's standards, the level of consumption registers quite low on the household scale, a municipal report from 1910 shows that the average consumption for domestic and industrial use rose to a mere 80 liters per day—a number that, combined with irrigation, fountains, and loss through leakage, would rise only slightly to the mandated 130 (Tello and Ostos 2012; Masjuan et al. 2008). Steadily but surely, the Sociedad General de Aguas de Barcelona grew to a virtual monopoly over Barcelona's water resources (Matés Barco 2019).

Yet, the Sociedad General de Aguas de Barcelona could not expand infrastructure at the rate necessary to accommodate the growing population. While industrialization and economic success of Barcelona did lower inequality from Renaissance levels, the development of an industrial proletariat created a stricter class structure that became ingrained in the spatial relationships between the urban poor and the rich. Higher areas received more costly water, while lower areas were generally cheaper, but the residential districts were mostly served by rooftop cisterns, many made of lead, which spread disease and offered little water in highly populated buildings. Thus, the poor suffered less water access, and the wealthy could avail themselves of amenities such as fountains, sanitation, and adequate piping throughout their districts (Masjuan et al. 2008).

The burden of water shortages, then, fell on the poor, particularly during the cholera and typhus epidemics that ravaged Barcelona in 1885 and 1914, respectively. Typhus, which spread as a result of bacteria-carrying flea, spread throughout the city but especially the poorer quarters where people struggled in piecework labor (Tello and Ostos 2012; Ostos and Tello 2014). Also, cholera ravaged areas that lacked proper sanitation and waste disposal. Both diseases, which claimed thousands of lives in a normal year, wreaked havoc in a city whose population grew far faster than its infrastructure could (Ruiz-Villaverde et al. 2010).

## 6.5 Water and Political Experimentation

During the 1920s, a fascist-inspired dictator named Miguel Primo de Rivera took charge of Spain in a coup d'etat, advancing a program of intensive urbanization. As people flocked to Barcelona, the dictatorship built a municipal showpiece called the Font Mágica, a majestic fountain flanked by waterfalls meant to show the imperial power of the city (Jaque Ovejero 2015) during exhibitions. At the same time, the exhibit showed off Barcelona's command over water resources and the implicit guarantee of fertility and bounty (Fig. 6.2).

The Sociedad General de Aguas de Barcelona had repelled efforts to municipalize the water resources in 1913, but the economic inequality and its impacts on the health and welfare of Barcelona's working poor caused a heavy backlash in the form of radical, left-wing political movements (Gemie 2011). From 1914 to 1936, the city's population doubled to 1.2 million residents. Slums grew, as more than 13% of the city's buildings lacked access to running water (Gorostiza et al. 2013). Density increased, particularly in the old city, causing sanitation to decline even as rents skyrocketed (March 2015).

The dictatorship banned the groups most responsible for protesting the state of daily life in Barcelona, including the Confederación Nacional del Trabajo (CNT). As a result, the demands for a 50% reduction in rents carried forward by the CNT transferred to decentralized advocacy groups assembled under the black and red flag of anarchy (Gorostiza et al. 2013). Because rents often included water access, these demands, along with the subsequent rent strikes, implicitly involved water and sanitation demands.

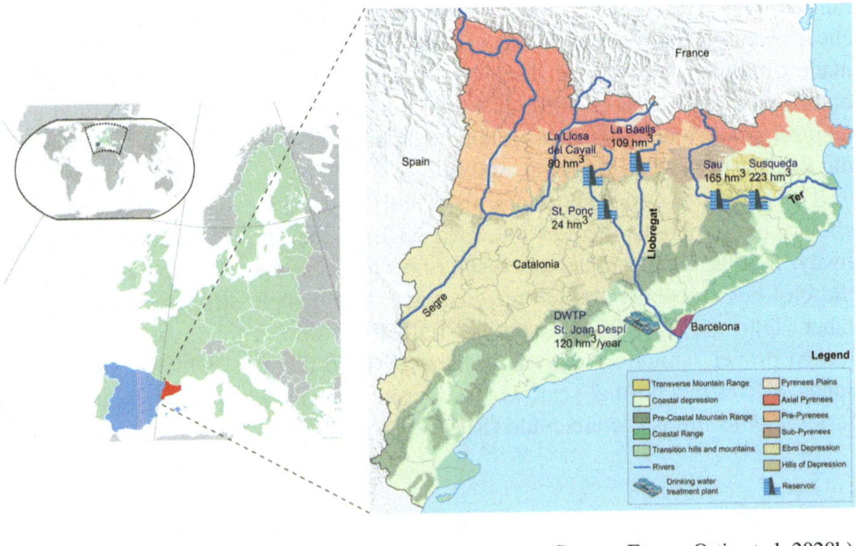


**Fig. 6.2** Location of Barcelona, major reservoirs, and rivers. (Source: Forero-Ortiz et al. 2020b)

In 1929, the global economy crashed and the Spanish monarchy fell, leading to the establishment of the Spanish Republic, which attempted to bring all the factions that dissented from the dictatorship together under the common ground of an elected, democratic form of government. Under the Republic, municipal authorities attempted to increase the number of water meters on properties and raise minimum water supply requirements for tenants of residential buildings; however, landlords resisted these efforts, and the Republic's water egalitarianism fell flat (March 2015).

In 1936, left-wing factions consolidated under the Popular Front model that had been adopted in France the year before as a coalition between Soviet-allied Communists and the older Socialist Party. This revolutionary, left-wing grouping won a heavily contentious election, leading to violent political clashes culminating in the invasion of Spain by Spanish General Francisco Franco. Amid the turmoil, the workers of the Sociedad General de Aguas de Barcelona organized and seized control over the consortium, ushering in a new phase of collectivized municipal water management, which saw some major advancements even amid the chaos of the Civil War (Gorositza 2019).

In one working class district served by a private water vendor with local spring water, an anarchist neighborhood organization expropriated the spring for the public water consortium, which developed infrastructure such as piping for area houses. In addition, the collectivized water consortium improved pumping systems at the Llobregat aquifer in efforts to increase supply to working class areas, while also bringing Sabadell and Terrassa, two nearby towns into Barcelona's supply system (Masjuan et al. 2008). From swimming pools to fountains to showers and more, Barcelona's revolutionary water company pleaded with the people to support its program on the basis of hygiene, insisting that the Catholic Church was the "most terrible enemy of water… conspiring against a world of pools and public baths," and that it, "inoculated into people the hate towards physical contact of the human flesh with water, air and light, as the highest moral idea" (Gorostiza et al. 2013).

However, efforts to split the water bill between a minimum quantity paid by the landlord and the rest paid by renters, combined with stressors brought on by the war, led many landlords to flee the city entirely. As a result, the collectivized Sociedad General de Aguas de Barcelona faced a crisis of payments. Eventually, 80% of water bills would go unpaid (Gorostiza et al. 2013). When air raids against Barcelona increased and Generalissimo Francisco Franco's forces closed in from the West, the city's energy and water supply faltered. Severe shortages led to massive cutbacks, while refugees thronged into the city. Once again, typhus ran rampant through the streets, as the dreams of revolutionary equality fell under the march of authoritarian dictatorship.

## 6.6 Dictatorship

While the dictatorship blamed the ailing state of Barcelona's infrastructure on the policies of the collective company, instead of the bombing, they credited management during the revolutionary period for expanding the franchise, and they kept the central modernizing policies set into place. In 1945, an extended drought set in, leading the restored Sociedad General de Aguas de Barcelona to make all efforts to keep the city placated by preventing restrictions. Finally, however, the company buckled to water pressures, imposing restrictions in 1953 after attempting to expand supply for 4 years (Santiago Díaz 2022).

The stress of the "prolonged drought" and the political backlash that it caused opened the way to further supply-side projects, such as the usage of Llobregat River surface water and the initiation of water transfers from the Ter River's Sau reservoir to the northeast of the city (Popartan 2020). At the same time, Franco regime tended to ignore the poor at best and to attempt to conceal them in shantytowns and districts far enough from the urban middle classes to provide an illusion of general wealth and civility. Yet, Franco froze wages amid inflation, while failing to promote rational housing policies. Meanwhile, his supporters in reactionary Spain cited water supply and sanitation as the cause of the Civil War and guffawed at the regime's responsibilities to the urban poor (Gorostiza et al. 2013).

As the attempts at wage controls and protectionism faltered throughout the 1950s, Franco increasingly liberalized Spain's economic program, leading to gains in economic growth but unemployment in the previously protected agricultural sector. As a result, the percentage of Spain employed in the agricultural sector shrunk from 50% in 1950 to just 15% in 1975 (Hamilton 2017). As a result of the large-scale migration of people from agricultural areas around Spain into Barcelona, shantytowns grew up around Spain. Without access to clean water or electricity, these slums came to house tens of thousands of people in unsanitary conditions, exhibiting the failure of the dictatorship to fulfill its promises of prosperity.

Meanwhile, in order to appease the laboring poor of the arid south, Franco built dams and canals, reclaiming vast areas of wetlands in an effort to generate a productivist system in which the economy could utilize all resources available to the fullest extent (Masjuan et al. 2008). Insisting that he would utilize Spain's water resources "to the last drop" in order to ensure agricultural production, by the 1960s, Franco regime fixated on water distribution schemes involving complex transfers and massive infrastructure projects (Swyngedouw 1999).

To resolve problems of water scarcity in Barcelona, during the late 1960s, the regime transferred surface waters from the Llobregat and Ter River systems to the municipal supply, expanding the supply of water. Shortly thereafter, however, the dictatorship put forward a new plan to satisfy the growth of the city by diverting the flow of the Ebro River from the lands of Aragon. Met with unusual protests in that province, the government canceled the Ebro water transfer plan. The initiation of large-scale water transfers became a kind of trademark for the Franco regime, which would ultimately backfire (Camprubí 2012).

Scholar Sarah Hamilton wrote that "Over the course of the lengthy dictatorship (1939–75), the Spanish state dammed rivers, drained wetlands, exterminated predator species, encouraged unregulated industrial and urban expansion, and expropriated vast swaths of public property for private use." "Such policies, intended to stimulate economic growth, had the secondary effect of centralizing control over resources and strengthening the power of the national government" (Hamilton

2017). During the 1960s, however, movements began to grow from landlords and external pressures to conserve wetlands and riparian areas, while environmental justice movements brewed in the slums.

Localized ecological issues led to the formation of more neighborhood associations, which then became launchpads for greater resistance, including clandestine political parties. In this way, ecology formed a basis for a democratic opposition not only to Franco but also to the regime he stood for. Some 200,000 people joined street protests called by environmental and anti-nuclear groups upon the death of the dictator in a mass movement that swept the authoritarian system from power and restored the Spanish Republic (Hamilton 2017).

## 6.7 Return of Municipalism

Barcelona's first City Council after municipal elections in 1979 dedicated itself to prioritizing the number and quality of green spaces, which had been neglected under the regime. Quality of life became the key focus of urban revitalization, as the burgeoning democracy sought to represent the demands that had faced fierce repression. Thus, the municipal government worked with neighborhood groups on the strategy of creating public space and urban infrastructure to better serve the public (Sürer 2018; Sanfeliu Arboix and Martín García 2017). This form of democratization also took place regarding water resources management.

While remaining driven by water infrastructure, the post-Franco system established a new water regime under the 1985 Water Law, which empowered River Basin Associations to develop management plans with the participation of local water users, irrigators, regional authorities, and hydroelectric companies (Bukowski 2007). These basin associations would coordinate through the National Hydrological Plan, which provided the central government with the ultimate say in water transfers and regional disputes while also democratizing the more technocratic tendencies (Frolova 2010). At the same time, the infrastructure-based program kept the 1897 Water Act's permitting practices, thus continuing the prior restrictions to quantities based on place and primary use.

The new Water Law was tied to the larger post-Franco administration of water resources and socio-natural systems throughout Spain, highlighted in Barcelona by the city's "ambitious program to bring the social and ecological benefits of urban green spaces to all parts of the city." Overall, this program involved supporting parks and forests, as well as encouraging stream restoration, urban agriculture, and ecological corridors (Anguelovski et al. 2018). In effect, the revival of the republican political system brought with it a kind of rebirth of municipal ecology last seen in Barcelona during the Civil War.

The early years of enthusiastic municipal renovation changed during the 1990s, however, as the symbolic decision to host the 1992 Olympic games in the city led to the mobilization of high finance necessary to develop massive Olympic parks and structures designed for commercial purposes rather than neighborhood democracy

(Sanfeliu Arboix and Martín García 2017). As the experience of tourism and life as a cosmopolitan global city set in, the municipality increasingly looked to developers to replace old neighborhood features and industrial sites with pricey new real estate and the architectural signifiers of post-industrial growth. Intriguingly, that same year, social justice protests erupted in some quarters of the city among residents who refused to pay the water fee, directing their affordability demands at the public administration (Popartan 2020). By the 2000s, the neighborhood-centered model of municipalism came into conflict with the so-called "Barcelona Model," a politically labile effort to couple social causes with urban development that often fell short of its principles (Charnock et al. 2021).

At the turn of the millennium, Barcelona became a critical flashpoint in a sweeping effort by the Spanish government to transfer water from "surplus basins" into basins with "structural deficits" (Saurí and Del Moral 2001). The main surplus once again came from the Ebro River in the province of Aragon, with its water destined for areas to the south such as Valencia and Murcia, as well as Barcelona, itself. Despite being on the receiving end of this redistribution of water resources, the population of Barcelona did not jibe with the overall project and protests spread throughout the country against what reminded many of the Franco regime's grand water schemes (Hernández-Mora et al. 2014).

To compensate for the inability to transfer the Ebro's waters to Barcelona's supply, planners attempted to administer a new program that would develop a technological solution to water scarcity—desalination.

With a mean decrease in fresh water of up to 11% for Barcelona by 2100, water concerns remain a serious priority (Forero-Ortiz et al. 2020a). Marketed as a commonsense solution to the city's water supply issues, desalination rapidly became a controversial topic. Activists compared the technocratic manipulation of water distribution involved in desalination plants to the water transfers under Franco, favoring instead more decentralized and community-oriented modes of water management. In addition, the need to emit fossil fuels in order to produce potable water from seawater during the desalination process brought many to view the effort to resolve water scarcity resulting from climate change–induced warming as a Pyrrhic victory that would merely feedback into the intensification of warming (Swyngedouw 2013).

In the late 2000s, a movement emerged to remunicipalize the private administration of water in the city in an effort to gain control of water resources. Started by professionals, engineers, and managers who sought to oppose corrupting aspects of public–private collaboration, the movement quickly spread to popular groups insisting, as in the days of the 1930s, on sanitation and water as a human right (Popartan 2020). Groups filed a lawsuit against a large water company for mismanagement, and though the suit was later won by the company in 2019, the effort helped to keep the topic in the public eye (Popartan et al. 2020).

With the mass popular movements that developed after the financial crisis of 2008 and the Arab Spring of 2011, the remunicipalization of water became a growing issue for society (Charnock et al. 2021). In 2015, the social movement platform Barcelona en Comú helped to assemble left-wing legislative lists that won the election in municipal government with remunicipalization as an important plank. However, due to legislative setbacks and splits over the question of Catalan independence, the effort to challenge the privatization of water resources in Barcelona failed to come to fruition.

## 6.8 Blue Infrastructure and SUDS

For Barcelona, climate change portends threats from reduced availability of freshwater resources and increasing urban flooding, as well as vanishing beaches and decreased water quality. Hazards such as flooding will increasingly threaten the extensive metro system in Barcelona's underground, requiring significant infrastructure investments (Forero-Ortiz et al. 2020a). Meanwhile, the estimated cost of flood damage comes to some 63 million euros unless the city develops comprehensive drainage systems to meet the challenges (Ortiz et al. 2020).

While much of the public discourse on water issues revolves around municipalization and the controversial implementation of desalination, according to data from the Climate Research Foundation, climate change will accelerate the entire water cycle in the region, leading not only to the intensification of droughts and heatwaves but also to that of storms (Forero-Ortiz et al. 2020b). The increase of maximum rainfall intensity in terms of frequency, rate, and timing will present challenges for Barcelona city planners who hope to keep rain out of the city as much as possible (Rodríguez et al. 2014). Flooding can create nuisances for traffic and pedestrians, but it can also lead to water quality problems, sewage overflows, and sanitation issues. Fortunately, the robust community participation at the neighborhood level that brings Barcelona so much renown also proved useful in adapting to climate-induced flooding.

In 2006, a new city master plan called the Comprehensive Plan for Barcelona's Sewerage System (PICABA06) in a forward-thinking effort considered first and foremost the potential impact of climate change in future rainfall scenarios. The plan further included climate-specific measures to mitigate harm, looking at social and economic, as well as environmental, damage that could take place. In particular, the comprehensive plan included analyses of urban resilience and economic efficiency using a long-term approach (Ortiz et al. 2020).

In the context of this long-term approach to harm mitigation through climate modeling in localized, resilience-based scenarios, the comprehensive plan pushed for more development of sustainable urban drainage systems (SUDS). These urban drainage systems have since been implemented in areas throughout Barcelona, particularly focusing on neighborhood engagement. In the Bon Pastor neighborhood, city officials called for the clearing of entire blocks of old, shoddily constructed residential buildings and their replacement through the neighborhood association with better developments. In the process, residents engaged in a collaborative process with water companies, consulting companies with specialization in SUDS around Spain, and the local government, meeting once or twice a week to discuss the design, implementation, and maintenance of blue-green infrastructure (Nóblega Carriquiry et al. 2020).

Here, the SUDS, which can include better drainage systems, cisterns to catch water, and gardens to infiltrate, filter, and percolate water into aquifers, depend on a kind of shared participation not only of the neighborhood but of the private sector and public partnerships. They also helped connect local managers to other managers around the country who also implemented SUDS, while developing inroads for residents uninvolved in the neighborhood association to join in broader community efforts. In structural terms, the SUDS developed in Bon Pastor represented a runaway success, removing strain from the traditional drainage systems entirely; however, in environmental terms, they also decreased the "heat-island" effect, helped recharge the aquifer, and enhanced the use of public space. Naturally, this led to further social integration at the neighborhood level and between residents and water managers (Nóblega Carriquiry et al. 2020).

## 6.9 Green Infrastructure and Biodiversity

The ideas put forward by the city's 2006 comprehensive plan and implemented through the SUDS infrastructure in neighborhoods such as Bon Pastor follow the same principles of "water-sensitive cities" discussed and promoted in the context of Melbourne, Australia. However, the Bon Pastor model also reveals another problem for Barcelona's urban administration, rampant sprawl. Barcelona's urban growth outward includes the claim of natural landscapes for residential development, creating fragmentation of green areas and problems of traffic and emissions. In 2010, the city attempted to confront this clear growth dilemma with the Barcelona Green Infrastructure and Biodiversity Plan 2020 (Parés et al. 2016).

According to Parés et al. (2016), "Restoring and enhancing [Green Infrastructure] through this plan provides the inhabitants of Barcelona with many ecosystem services such as air purification, noise reduction, regulation of urban climate and temperature, reduction in energy consumption and $CO_2$ emissions, water cycle regulation, recreation, improvement in mental health and general well-being." Green infrastructure also offers greater connectivity between green areas, cracking the problem of fragmentation (Langemeyer and Baró 2021).

Through a participatory approach, the municipal government developed a plan including political representatives as well as scholars, environmental organizations, and a variety of nongovernmental institutions to promote strategies for improving green infrastructure. These efforts produced renewed efforts to maintain parks and other green spaces in a way that reduces water use, as well as that ensures sustainable soil use and pest control (Camps-Calvet et al. 2016). The cooperative also developed communications strategies to enhance understanding and engagement in supporting green infrastructures toward health, fun, and community leadership (De Bellis et al. 2015).

At the same time, the City of Barcelona has come to the harsh reality that its resources must be split between green infrastructure and maintenance of existing water infrastructure—fixing pipes, implementing smart irrigation, and sustainably drawing on groundwater. This means that, rather than generating new green infrastructure, the city's plans increasingly focus on maintenance. However, through the 2012 "Plan of Vacant Urban Lots," local NGOs now work to convert vacant lots into green spaces that can serve as community gardens and other means of community and cultural belonging (Orduña-Giró and Jacquot 2015).

At the same time, some ambitious alternative plans include the conversion of larger agricultural areas in the metropolitan area into more intensive farming on smaller plots, rendering areas of the fields either back to scrublands or into forestlands. This effort would include agroforestry and mixed use to maintain the productivity of production while also meeting socioecological needs at the same time (Basnou et al. 2020).

While the growth of green infrastructure in Barcelona benefits flood management, there is an increasing body of research that shows that the expansion of green space in several historically underserved districts around Barcelona likely led to "an above average increase for their district in residents with a bachelor's degree or higher (except in Princep de Girona Garden), residents from the Global North (except Diagonal Mar10), household income or home sale values, and a decrease in the population over 65 living alone" (Anguelovski et al. 2018). Some underserved communities did gain increased access to green spaces but remained less connected to the city and had lower-quality housing with higher threats of foreclosure (García-Lamarca et al. 2020). In the case of Barcelona, green infrastructure is understood to be integrated within gentrifying tendencies, rather than leading those tendencies, but importantly, such a path is not economically determined and could be mitigated with careful planning (Charnock et al. 2021).

## 6.10 Conclusions

A number of conclusions can be drawn about the Barcelona model of water management in recent years. From its adaptation to flood hazards to its inclusive approach to neighborhood engagement, Barcelona has offered a beacon of hope to large metropolises throughout the world. However, with that hope comes a price as Barcelona's efforts to enhance blue-green infrastructure through local involvement tend to contribute to an increase in property values and a demographic shift toward wealthier, more-educated residents. While this "paradox" may present a wicked problem in terms of complex, adaptive social-ecological-technological systems, municipalist efforts to enhance spatial equity could offer better results.

Barcelona's promise of municipal water management stems from prerevolutionary and revolutionary efforts to gain equity for residents of poorer neighborhoods. Such efforts tended to empower neighborhoods and standardize fees, bringing access to underserved communities and lowering landlords' ability to control affordability. The efficacy of such reforms was indicated by the Franco regime's maintenance of the new, more hygienic codes, in spite of the dictator's efforts to distance himself from the prior government.

Following Franco, whose neglectful approach to slums led to a decline in urban hygiene, Barcelona's city leaders drew on the active social and ecological movement coming out of the neighborhoods, which in some ways kept alive the spirit of the revolutionary period. While attempts to increase green spaces distributed throughout the city and activate community engagement to manage blue-green infrastructure proved successful, it also brought with it ongoing problems. The development of megaprojects and the implications of green infrastructure on spatial–economic conditions, along with the stalled condition of the municipalist water movement, show perhaps that the Barcelona model has brought financial success and health improvements, but despite efforts at repurposing toward municipalism, it has a long way to go to provide support for underserved communities (Charnock et al. 2021).

Yet, the success of Barcelona's blue-green infrastructure in mitigating flood risk can hardly be overlooked as one of the more exemplary models of urban water and resilience to climate change surveyed in the present volume. One of the lessons to take from Barcelona is that an active and engaged community is the fundamental necessity for adaptive Social-Ecological-Technological Systems. That engagement tends to come from a feeling of empowerment, which relies on opportunities to work with and through local governing structures in a collaborative framework. At the same time, the existence of Barcelona's strong neighborhood organization makes it a somewhat novel example and indicates a difficult structural baseline from which other cities might build.

## References

Alba DM, Moyà-Solà S, Robles JM, Galindo J (2012) Brief communication: the oldest pliopithecid record in the Iberian Peninsula based on new material from the Vallès-Penedès Basin. Am J Phys Anthropol 147(1):135–140

Anguelovski I, Connolly JJ, Masip L, Pearsall H (2018) Assessing green gentrification in historically disenfranchised neighborhoods: a longitudinal and spatial analysis of Barcelona. Urban Geogr 39(3):458–491

Basnou C, Baró F, Langemeyer J, Castell C, Dalmases C, Pino J (2020) Advancing the green infrastructure approach in the province of Barcelona: integrating biodiversity, ecosystem functions and services into landscape planning. Urban For Urban Green 55:126797

Bukowski J (2007) Spanish water policy and the national hydrological plan: an advocacy coalition approach to policy change. South Eur Soc Politics 12(1):39–57

Camps-Calvet M, Langemeyer J, Calvet-Mir L, Gómez-Baggethun E (2016) Ecosystem services provided by urban gardens in Barcelona, Spain: insights for policy and planning. Environ Sci Pol 62:14–23

Canérot J (2017) The pull apart-type Tardets-Mauléon Basin, a key to understand the formation of the Pyrenees. BSGF-Earth Sci Bull 188(6):35

Charnock G, March H, Ribera-Fumaz R (2021) From smart to rebel city? Worlding, provincialising and the Barcelona model. Urban Stud 58(3):581–600

Camprubí L (2012) 'Frankie the frog': the total transformation of a river basin as 'totalitarian'technology (Spain, 1946–1961). Endeavour 36(1):23–31
De Bellis Y, Santos A, Tosics I, Davies C, Hansen R, Rall E, ..., Lafortezza R (2015) Green infrastructure planning and implementation. The status of European green space planning and implementation based on an analysis of selected European cityregions. GREEN SURGE PROJECT, Deliverable 5
Enrich Pericás R (2022) Aproximació als sistemes hidràulics medievals: Una anàlisi preliminar als territoris de Catalunya, la Comunitat Valenciana i les Illes Balears
Forero-Ortiz E, Martínez-Gomariz E, Cañas Porcuna M, Locatelli L, Russo B (2020a) Flood risk assessment in an underground railway system under the impact of climate change—a case study of the Barcelona metro. Sustainability 12(13):5291
Forero-Ortiz E, Martínez-Gomariz E, Monjo R (2020b) Climate change implications for water availability: a case study of Barcelona City. Sustainability 12(5):1779
Frolova M (2010) Landscapes, water policy and the evolution of discourses on hydropower in Spain. Landsc Res 35(2):235–257
García-Lamarca M, Connolly J, Anguelovski I (2020) Green gentrification and displacement in Barcelona. In: Housing displacement. Routledge, pp 156–170
Gemie S (2011) Anarchism and the City: revolution and counter-revolution in Barcelona, 1898–1937. Anarch Stud 19(1):109
Gorostiza S, March H, Sauri D (2013) Servicing customers in revolutionary times: the experience of the collectivized Barcelona water company during the Spanish civil war. Antipode 45(4):908–925
Gorositza S (2019) Critical networks. In: Laakkonen S, McNeill JR, Tucker RP, Vuorisalo T (eds) The resilient city in world war II: urban environmental histories. Springer
Guardia M, Rosselló M, Garriga S (2014) Barcelona's water supply, 1867–1967: the transition to a modern system. Urban Hist 41(3):415–434
Hamilton SR (2017) Environmental change and protest in Franco's Spain, 1939–1975. Environ Hist 22(2):57–281
Hernández-Mora N, del Moral Ituarte L, La-Roca F, La Calle A, Schmidt G (2014) Interbasin water transfers in Spain: interregional conflicts and governance responses. In: Globalized water: a question of governance, pp 175–194
Instituto Geográfico Nacional (2015) Geology and tectonics of the Catalan coastal range. Gobierno de España, Madrid. https://www.ign.es/web/resources/sismologia/tproximos/sismotectonica/pag_sismotectonicas/costerocatalana_en.html
Jaque Ovejero A (2015) Mies en el sótano. El Pabellón dé Barcelona como ensamblajes de lo social. Doctoral dissertation, Arquitectura
Kreibich H, Van Loon AF, Schröter K et al (2022) The challenge of unprecedented floods and droughts in risk management. Nature 608:80–86. https://doi.org/10.1038/s41586-022-04917-5
Lagabrielle Y, Labaume P, de Saint Blanquat M (2010) Mantle exhumation, crustal denudation, and gravity tectonics during cretaceous rifting in the Pyrenean realm (SW Europe): insights from the geological setting of the lherzolite bodies. Tectonics 29(4):1–26
Langemeyer J, Baró F (2021) Nature-based solutions as nodes of green-blue infrastructure networks: a cross-scale, co-creation approach. Nat Based Solut 1:100006
March H (2015) Taming, controlling and metabolizing flows: water and the urbanization process of Barcelona and Madrid (1850–2012). Eur Urban Reg Stud 22(4):350–367
Masjuan E, March H, Domene E, Saurí D (2008) Conflicts and struggles over urban water cycles: the case of Barcelona 1880–2004. Tijdschr Econ Soc Geogr 99(4):426–439
Matés Barco JM (2019) El abastecimiento de agua a Barcelona (1850–1939): origen y desarrollo de las compañías privadas. Historia Contemporánea 59:161–194
Nóblega Carriquiry A, Sauri D, March H (2020) Community involvement in the implementation of sustainable urban drainage systems (SUDSs): the case of bon pastor, Barcelona. Sustainability 12(2):510

Orduña-Giró P, Jacquot S (2015) The participatory production of temporary public spaces in times of crisis. The 'Pla Buits' project in Barcelona

Orengo HA, Miró i Alaix C (2013) Reconsidering the water system of Roman Barcino (Barcelona) from supply to discharge. Water Hist 5:243–266

Ortiz A, Velasco MJ, Esbri O, Medina V, Russo B (2020) The economic impact of climate change on urban drainage master planning in Barcelona. Sustainability 13(1):71

Ostos JR, Tello E (2014) A long-term view of water consumption in Barcelona (1860–2011): from deprivation to abundance and eco-efficiency? Water Int 39(5):587–605

Parés M, Rull C, Rivero M (2016) Case Study 9 Spain: developing the Barcelona green infrastructure and biodiversity plan. Nature-Based solutions to address global societal challenges. In: Cohen-Shacham E, Walters G, Janzen C, Maginnis S (eds). International Union for Conservation of Nature (IUCN). Gland, Switzerland.

Popartan LA (2020) The socio-cognitive dimension of water: the case of politicisation of water in Barcelona. Dissertation. University of Girona

Popartan LA, Ungureanu C, Velicu I, Amores MJ, Poch M (2020) Splitting urban waters: the politicisation of water in Barcelona between populism and anti-populism. Antipode 52(5):1413–1433

Popartan LA, Ungureanu C (2022) The political ecology of water memory: contending narratives of past hydraulic infrastructures in Barcelona (2015–2021). Polit Geogr 96:102596

Rodríguez R, Navarro X, Casas MC, Ribalaygua J, Russo B, Pouget L, Redaño A (2014) Influence of climate change on IDF curves for the metropolitan area of Barcelona (Spain). Int J Climatol 34(3):643–654

Ruiz-Villaverde A, García-Rubio MA, González-Gómez F (2010) Analysis of urban water management in historical perspective: evidence for the Spanish case. Int J Water Resour Dev 26(4):653–674

Salgot M, Angelakis AN (2019) The historical development of water supply technologies in Barcelona, Spain. In: Evolution of water supply through the millennia. IWA Publishing, London

Sanfeliu Arboix IR, Martín García E (2017) Public space in Barcelona (1992–2017): evolution and case studies. In: IOP conference series: materials science and engineering, vol 245. Institute of Physics (IOP), pp 1–11

Santiago Díaz G (2022) Culpa de la guerra, culpa de Franco. La hambruna española en la Andalucía Oriental rural de posguerra (1939–1953)

Saurí D, Del Moral L (2001) Recent developments in Spanish water policy. Alternatives and conflicts at the end of the hydraulic age. Geoforum 32(3):351–362

Sürer İ (2018) The emergence of public space in Barcelona throughout 20th century. Doctoral dissertation, Universitat Pompeu Fabra

Swyngedouw E (1999) Modernity and hybridity: nature, regeneracionismo, and the production of the Spanish waterscape, 1890–1930. Ann Assoc Am Geogr 89(3):443–465

Swyngedouw E (2013) Into the sea: desalination as hydro-social fix in Spain. Ann Assoc Am Geogr 103(2):261–270

Tello E, Ostos JR (2012) Water consumption in Barcelona and its regional environmental imprint: a long-term history (1717–2008). Reg Environ Chang 12:347–361

Thompson JKJ (2002) A distinctive industrialization: cotton in Barcelona 1728–1832. Cambridge University Press, Cambridge

Tinoco Mosteiro H (2022) Les sitges de cereals de l'edat del Ferro a la conca del Llobregat. Estudi morfomètric i qualitatiu de les estructures de conservació d'excedents

Vallvé MB, de la Peña JCR, Vide JM (1998) Instrumental Calibration of climate data. A methodological approach to annual resolution in reference to Barcelona's precipitation (1521–1989). Investigaciones Geográficas (20), 99

Vázquez FM (2023) Sobre la implantación del arquetipo forense imperial en Hispania. La "provincialización" como concepto y modelo de difusión. Lucentum

Vicente M (2000) Artisans and work in a Barcelona cotton factory (1770–1816). Int Rev Soc Hist 45(1):1–23

# Chapter 7
# Lagos, Nigeria

**Abstract** Lagos, Nigeria, is one of the most economically successful cities in Africa, providing a crucial hub for the movement of goods while harboring profitable commercial enterprises in multiple sectors. Yet, as a result of its development as a colonial center in an ecologically precarious area subject to flooding, as well as its rapid-pace urbanization as a major regional population center, the megacity has inherited many problems involving water management. This chapter examines the role of water resources in the growth of Lagos, including the substitution of wetlands for urban districts amid crises of disease and inequality. The plight of informal settlements is discussed apropos the disparate burden borne by women in particular as well as the efforts of nongovernmental organizations, academic institutions, and the state to address the uneven distribution of water hazards and resources. Analyses are foregrounded that view water issues as inextricably tied to the spatial inequality of housing and a lack of investment in poor communities, indicating holistic solutions that address not only the typical hydrologic approach of "moving water out of the city" but also helping communities take root and address the problems of water quality and quantity with localized and decentralized means coordinated across multiple scales.

**Keywords** Urbanization · Spatial inequality · Social networks · Human rights · Flooding

## 7.1 Introduction

Nineteen out of Africa's 90 coastal cities have more than a million residents, and Lagos is the largest of them all. A megacity with a population in excess of ten million residents, Lagos is a crucial economic hub for sub-Saharan Africa and the world economy. According to the Climate Change Vulnerability Index, however, Lagos is one of the top ten "high-risk" cities facing climate hazards (Adelekan 2016). Other analyses put Lagos in the top 15 cities most exposed to climate change by 2070 (Oshodi 2013). Part of the reason for Lagos' high risk lies in systemic


H. Chang, A. R. Ross, *Climate Change, Urbanization, and Water Resources*, https://doi.org/10.1007/978-3-031-49631-8_7

problems of economic inequality and its relationship to land and power. For this reason, although Nigeria is taking considered steps toward integrating climate vulnerability into policy and embracing international bodies dedicated to climate awareness and adaptation, scholars and activists insist that a holistic human rights framework is the only hope for a fair and equitable approach to climate resilience in Lagos (Ajibade and McBean 2014).

Like other colonial cities, Lagos was produced not by a natural historical progression but by the exigencies of economic production through imperial systems. People moved into the area not because the geographic situation of Lagos offered a particularly hospitable environment but because imperial governors and crises elsewhere developed limited infrastructure to facilitate the export of goods. From its inception, then, populations in the low-lying area of Lagos have been vulnerable to floods, disease, saltwater intrusion, and subsidence.

Yet, as a result of colonial policies privileging colonial administrators and postcolonial policies furthering the systemic privilege, the class would continue to determine geographic exposures to climate hazards. Also, due to political and economic instability, including the debilitating Civil War and multiple military dictatorships, a broken line of attempted infrastructural improvements pertaining to water supply and use has been riddled with administrative and bureaucratic failure (Gandy 2006). However, as Nigeria moves toward a political philosophy of improved collaboration, integration, and inclusion, new policy directions are emerging from a generation of scholars and activists working to instill human rights–based approaches to resilience through equalizing measures that range from the place of women in the household to the dire need for housing improvements in Lagos' slums (Bollaert et al. 2023; Ajibade and McBean 2014).

## 7.2 Climate and Physical Geography

The coast of West Africa stretches 2000 miles from Gibraltar through the Mediterranean climate of Morocco in the Western Maghreb and down past the arid Sahara and Sahel band to Senegal, where it winds another 1300 miles southward through more lush climes before cutting east toward the Gulf of Guinea at the border of Liberia and the Ivory Coast. A final 800 miles along the coast resolves in the Gulf of Guinea. The western bight of the Gulf of Guinea, called the Bight of Benin, proved a critical strategic location for trade and war. No location along this vast and diverse coastline has become richer, more populous, and perhaps more imperilled by climate change than Lagos, Nigeria.

At the boundary of the Atlantic and the African Plate, Lagos lies atop a huge sedimentary basin that extends from Togo to Benin and ends at southwest Nigeria's Okitipupa ridge. Formed after the separation of the African–South American continents, this composition of different sedimentary rocks rises from −9 m to 77 m throughout Lagos. As a result, what is now Lagos City sits on land that can easily

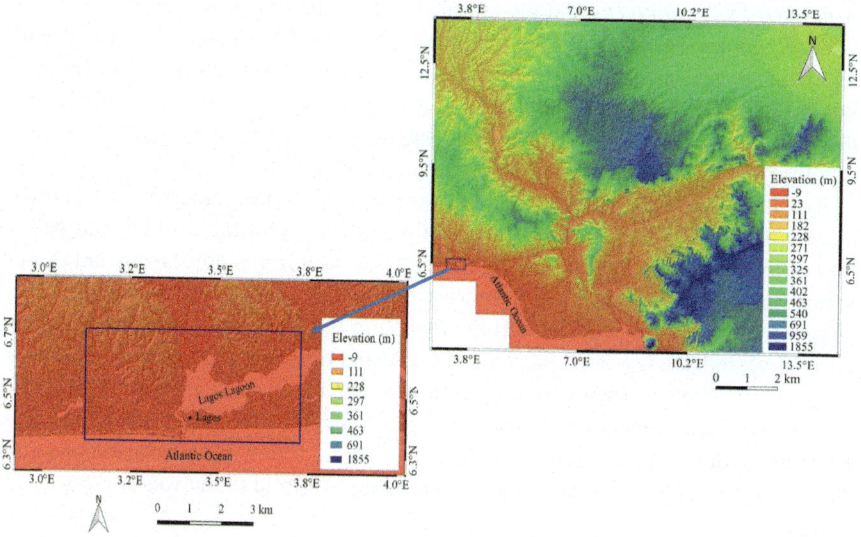


**Fig. 7.1** Elevation map for Lagos, Nigeria, including a Lagos Lagoon pullout. (Source: Ikuemonisan et al. 2020)

subside due to saltwater intrusion in aquifers, erosion, and flooding, as well as anthropogenic causes (Ikuemonisan et al. 2020) (Fig. 7.1).

The coastal belt of Lagos comprises vital wetland ecosystems within the upstream basins of the Oshun, Ogun, and Yewa rivers (Agboola et al. 2016). Defined as "areas of marsh, fen, peatland or water, whether natural or artificial, permanent or temporary, with water that is static or flowing, fresh, brackish or salt including areas of marine water the depth of which at low tide does not exceed six meters," wetlands offer critical sites for biodiversity around the world, particularly for migratory species (Wasserman and Dalu 2022). Among the most crucial wetlands in Nigeria, the Lagos coastal floodplains and wetland belt areas offer rich soils, fish habitat, and resources for human populations.

Water covers about 22% of the surface of Lagos state, and most of its wetland areas are low-lying, undulating, and flat, ranging from a mix of clay and sandy soil at sea level to more loamy soil about 50 feet above sea level (Oteri and Ayeni 2016). The major water bodies—the Ogun River and Lagos Lagoon—drain broader regions of marsh forest and swamplands. Situated amid the inter-tropical convergence zone (ITCZ) just north of the equator, Lagos typically benefits from a hot, low-pressure climate region fed by dry air sweeping in from the mid-latitude ridge that perpetuates the Sahara Desert. While the mean relative humidity lies at around 81.65%, there are alternative wet and dry seasons that pivot from the former to the latter in October and vice versa in April (Tejuoso 2006).

Extending from the hills of the agricultural lands of Oyo State and passing 186 miles through Southwestern Nigeria before emptying into the Lagos Lagoon, the Ogun river drains a lush riparian habitat for threatened species such as the

*Arnoldichthys spilopterus* and the kéwel (Ikomi et al. 1997). Fed by the Ogun, the Lagos Lagoon is the largest of ten major lagoons in the Lagos area, extending nearly 2500 square miles from the eastern part of the city into a strait that connects to Lekki Lagoon at the furthest reaches of Lagos State (Obiefuna et al. 2013). Indeed, the name Lagos derives from the Portuguese word for "Lakes," assigned to the location by explorers and merchants in the sixteenth century.

Traditional indigenous customs and practices continue today in the form of coastal fishing communities that utilize "the knowledge of the ecology and behaviour of fishes, navigation, harvesting methods, as well as the processing and preservation methods employed for centuries." After harvesting catfish, cooperatives largely maintained by women feed the fish palm oil, fruit, and water hyacinth, and once harvested, the fish are typically smoked using firewood and charcoal (Kolawole et al. 2010). The ancestors of the women who engage in these traditions may have practiced the religion popular among the Yoruba peoples who prevailed in the Lagos area during first contact with Portuguese navigators, followed by invasion and conquest by the people of the neighboring Kingdom of Benin to the west.

## 7.3 History of Sanitation and Development in Lagos

Six Yoruba subgroups—the Ilaje, Awori, and Ijebu—continue to populate the region, known as Yorubaland, that traverses some 55,000 square miles around Lagos, including parts of Togo and Benin (Olukoju 2017). While the area maintained some strategic and economic utility through the period of Benin's civil war in the late seventeenth century, the lucrative ivory trade largely came through the Niger River Delta, while the Kingdom's cultural and political power concentrated to the west of Lagos (Girshick and Thornton 2001). It was not until the height of the Atlantic slave trade during the early nineteenth century that Lagos took a prominent place in the economic geography of Africa, growing from 5000 to 20,000 in the three decades before 1810 (Law 1978).

After the British Empire intervened in 1861 to colonize Lagos and end its external slave trade, the city's economic production continued to founder amid colonial policy. Attached to a variety of different colonized regions at different times over most of the ensuing three decades, the early death rate of 30% for infants rendered the area considerably less attractive for investment (Van Tol 2007). While the British noted the deficiency in urban planning—particularly street layout and sanitation—they did not want to spend necessary resources, leaving Lagosians with the majority of the burden (Gandy 2006). The prevalence of water offered economic blessings in the form of transportation infrastructure and fishing, along with the looming curse of disease.

While histories of colonialism tend to present the imperial power as a virtually omnipotent force over the planning and design of new urban settlements, recent research shows that indigenous conceptions of development often differed from and even won out against the British ones." In spite of the fact that in the course of

British rule the Lagosians were subject to several colonial programmes relating to urban form and sanitation," writes one scholar, "their contribution to the island's layout was remarkable" (Bigon 2005). Whereas the British focal sites of socioeconomic interest—the Marina and Broad Street—were regulated, ordered, and designed according to European ideals of rational commerce, the Yoruba areas left to a growing influx of migrants followed different, more traditional models of family compounds and natural streets. Although the racial segregation of Lagos was a gradual, haphazard process, partly as a result of the neglect by its colonial overseers, the layout of water pipes was sparse, serving the white population more (Bigon 2009; Oluwasegun 2017).

Public Works and Sanitation Departments sprung up immediately after Lagos gained relative colonial independence in 1887, followed by a Board of Health in 1899 (Campus 2009). While these formal institutions did create a night soil removal service using evening tram runs along with other developments, they served more to avoid costly disorganization than to systematically resolve obvious problems. Much of the everyday sanitation work was left to the growing population, which autonomously developed infrastructure and practices for the disposal of waste. The gradual reclamation of wetlands and other mosquito breeding grounds came about only as population growth rendered such approaches unavoidable (Uwa 2018).

After the construction of a railway facilitated the transport of large stones used to build an official port, Lagos became the capital of a consolidated Colony and Protectorate of Southern Nigeria. Further consolidation of the Northern and the Southern Protectorates of Nigeria enabled broader development as a "First-Class Township" in accordance with the colonial authority, However, as Hopkins (1966) wrote, "It is one of the ironies of West African history that the rise of 'legitimate commerce,' the trade which was intended to supplant the sale of slaves, should have helped to perpetuate internal slavery during the second half of the nineteenth century." Lagos' position along the coast with access to the interior made it an increasingly lucrative capitol for the British Empire, yet the complicated and disjointed course of exploitative development led to a continued neglect of the coastal urban area and the people of the interior.

Meanwhile, efforts to impose a water rate met with resistance based on the inequitable distribution of revenues. A new party called the Nigerian National Democratic Party emerged, in part, to resolve water and sanitation issues in 1923, and outbreaks of bubonic plague beginning the next year added impetus to the push (Gandy 2006). However, the global economic depression curtailed most efforts outside of slum clearances, which contributed, in turn, to the development of tribal unions as political organizations to encourage solidarity and public representation.

Colonial administrators habitually blamed water and sanitation issues on unregulated settlements undertaken by natives, rather than a lack of investment or leadership, and authorities ensured that the brunt of the impacts of water shortages was felt by the poorest residents. As well, political corruption sabotaged infrastructure developments, as with the subversion of a 1956 underground sewage proposal by a group with business interests in the night soil collection business. By the time of independence in 1960, Lagos had grown immensely to become a complex city

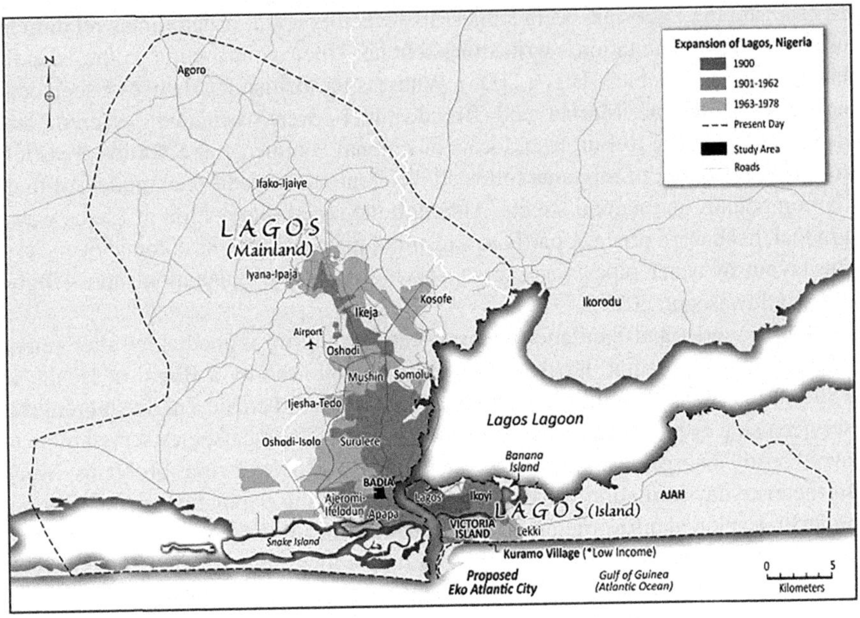


**Fig. 7.2** Map of expansion of the City of Lagos. (Source: Ajibade and McBean 2014)

developed largely by colonial interests with a large manufacturing sector numbering around 200 factories that employed nearly 30% of the male workforce as skilled laborers, and yet "underdeveloped" as a result of a general policy of extensive exploitation left water services largely forsaken (Gandy 2006) (Fig. 7.2).

## 7.4 Postcolonial Policy

Obtaining independence in 1960, Nigeria became an immediate cultural hub for a developing postcolonial world. However, the fledgling municipal government of Lagos employed, by the middle of the decade, only 30 planners, and one engineer ran the city's entire water distribution system. Just 10% of the metropolitan population drew water directly from the water system, while the rest used a mix of wells, polluted creeks, sparsely distributed standpipes, and communal taps—a proportion that would remain relatively consistent up to the present day (Gandy 2006). With the colonial model of slum clearances and restrictive access to land, the postcolonial state quickly reproduced systemic inequalities (Bigon 2008).

Formerly, colonial districts that received services simply turned into governmental residential areas, entrenching the spatial distance between rich and poor, while the country's complex national, tribal, and ethnic composition involved intense hegemonic contradictions. After the terrible 1967 Civil War, in which over a million people died, the military dictatorship replaced Lagos' city council with a new governmental entity, disrupting regional development while an influx of migrants came into the metropolitan area. Along with aggressive slum clearances, the migration waves altered the demographics of different municipal areas, leading to the erosion of tribal unions and the traditional, autonomous modes of sanitation services in the city. At the same time, from the 1970s onward, a general lack of water and other infrastructure contributed to the crisis of Lagos' industrial sector, ultimately contributing to its disintegration and replacement with massive slums (Gandy 2005).

When the military dictatorship fell in 1979 to civilian rule, efforts to curtail the water and housing crisis arose, like the 1980 Master Plan for Metropolitan Lagos, but it would attain only 10% of its housing goals despite international backing (Smith 1980; Gandy 2006). Civilian Governor Lateef Jakande built new micro- and mini-waterworks throughout Lagos, and the construction of new dams with the expansion of the Iju waterworks brought more potable water to residents (Omogunloye and Ayeni 2012). Yet, lack of bill payment and a high rate of infrastructural wear and tear plagued the system, and at its height, Lagosian water services only flowed "at a reduced level of service" to 47 percent of the population (Olukoju 2017).

Meanwhile, the substitution of industrial growth with a petro-economy enriched some and caused the growth of some infrastructural projects, which was then cut short by the oil crisis of the early 1980s as revenue fell from US $26 billion to $6 billion in just 6 years (Ibem 2011). With hyperinflation devastating the whole economy, a new military dictatorship took power and in 1986 supplemented a growing foreign debt with structural adjustment, further clamping down on the scanty public financing of municipal infrastructure.

From 1986 to 2002, a massive reclamation effort went into effect to make Lagos habitable for a fast-growing population. Urban area increased by 42%, while forest and grassland fell by more than 40% each—a precipitous drop of more than one-third over the course of just over 15 years (Twumasi et al. 2020). Along with rising lagoon areas, governmental failures, and fluvial flooding, this land reclamation is a top contributor to flood intensity and frequency in Lagos. Furthermore, the staggering loss of biodiversity due to clearances permanently damages the regional ecosystem, thus compromising endangered species habitat and the ability to counteract increased pollution from waste dumps, industrial emissions, and automobile traffic (Adewole 2009; Manzoor and Sharma 2019).

As Matthew Gandy (2006) writes, since the 1990s, "A self-service city has emerged in which little is expected from municipal government and much social and economic life is founded on the spontaneous outcome of local negotiations." This form of spatially fragmented governance and polarized society, in which residents provide their own energy, security, and water, or rely on expensive intermediaries with fluctuating services—an informal economy for potable water that employs a range of actors from underground crime networks to children seeking to supplement their parents' wage or provide for themselves. In a city that handles some 80% of Nigeria's sea transport and contributes more than a third of the country's total GDP, Lagos' climate hazards threaten the most vulnerable populations in Lagos as well as the entire country's economic production, which has significant global implications (Elias and Omojola 2015).

## 7.5 Climate Impacts in Lagos

Climate change poses serious risks to the already weak water infrastructure in Africa's most populous city. Lagos lies in a "vulnerable ecosystem subject to severe anthropogenic and natural hazards, such as sea level rise, land subsidence, flooding and coastal erosion and salinization of groundwater" (Yusuf et al. 2018). As the population of Lagos continues to grow, more people are filling its 100 slums that contain 70% of the population in highly vulnerable areas with little access to infrastructure or resources (Adelekan 2010). Thus, lawmakers, regulators, and administrators are met with a double bind: The municipal infrastructure is scanty, but further development could increase anthropogenic risks without careful planning.

With Lagos expanding from 200 to 1140 square kilometers over the course of just 40 years, rapid metropolitan development often came about in a spontaneous and unplanned fashion, necessitating the clearing of mangrove forests and reclamation of wetlands. Some coastal areas lost between 38% and 100% from 1986 to 2006 (Yusuf et al. 2018). Yet, the mangrove forests once served as a buffer for storm surges and the kind of intense storms that occur in the inter-tropical convergence zone (ITCZ). This is significant because the ITCZ's sinuosity and variability cause uncertainty in the timing of Lagos' rainfall (Opeyemi 2017). With higher temperatures, this warm, humid equatorial air will expand to hold a larger volume of water vapor, causing less but more intense storms. As glacial ice melts and the sea expands based on rising sea temperatures, sea level rise will imperil Lagos even as storms worsen (Fashae, and Onafeso 2011).

Such a phenomenon has already been observed by Adelekan (2010). From 1971 to 1995, Lagos saw more rainy days but less rain per rain day than in the ensuing decade. The rising intensity of storms was further indicated by public reporting of flooding between 2002 and 2006, with the percentage of people reporting rising from 54% to 71% in just 4 years. In response, policymakers produced Operation De-flood Lagos to upgrade some drainage infrastructure and clear drainage channels; however, such plans cannot keep pace with the hazards of population increase and climate change (Adelekan 2010).

Studies ranking projected exposure to sea level rise in urban areas around the world place Lagos within the top 50. As a result of Lagos' exposure and the number of vulnerable people dwelling in hazard-prone areas, Nigeria is considered among the 11 states with a port city imperiled by rising seas. Sea level rise can exacerbate flooding from storm surges and intense storms and lead to increased salinity of local water sources—a serious hazard considering Lagos' water systems. By 2070, without intervention, the number of people exposed to such hazards will have increased eight times over (Adelekan 2010).

Partly as a result of sea level rise, but also due to the increased intensity of storms, Lagos will see worsened effects of flooding. Of 136 cities analyzed with respect to 2070 climate conditions at the current rate of global emissions, Lagos was ranked 15th in people exposed to flood risk (Oshodi 2013). Flooding can occur multiple times in a single year, with incidents often causing knee-high or waist-high waters that can destroy homes. Such risks create poverty cycles, wherein poor people develop informal settlements and "slums" in vulnerable areas without adequate drainage and with little infrastructure (Adelekan 2010).

As a result of population growth and the increased flooding preceding the decline in wetlands and the rise of urbanization, erosion and well-tapping cause land subsidence, which worsens floods by dropping the land closer to or even below sea level. Each year, some 45% of Lagos is subject to seasonal flooding and erosion, such that subsidence emerges as a crucial problem. Subsidence causes damage to infrastructure and cracks in foundations, which can further exacerbate subsidence as foundations fail. Importantly, subsidence gets worse by proximity to the coastline, thus elevating exposure to poorer communities living in slums throughout those areas (Ikuemonisan et al. 2020).

## 7.6 How Climate Impacts the Poor

In the average Lagos slum, at least four people dwell in a one-room home in a house shared by 8–10 different households. Such areas are often anywhere from 40% to 90% developed, with buildings and roads that make the ground more impervious to rain infiltration, thus increasing the risk of flooding. Since many buildings in slums are constructed using waste to fortify foundations (called waste-filling), flooding can cause the breakdown of waste-filling, failing of foundations, and further pollution of groundwater (Ajibade and McBean 2014). Without safe sanitation systems, flooding can cause not only the destruction of houses and death but also areal contamination with sewage and waste from uncovered sewers and inadequate pipes that can remain in water sources such as wells, as well as food sources, leading to cholera and other diseases (Ayeni 2014).

Higher temperatures will exacerbate the threat of diseases, along with increasing pressures on the elderly and young. The hotter weather, along with shifts in rainfall intensity and timing, further impacts fish populations, which impacts the population and economy. According to one study, 100% of respondents reported the prevalence of diseases such as malaria and income loss from sickness due to flooding (Abiodun 2021). With little to do but rebuild or relocate, populations packed into an area of up to 40,000 people per square kilometer are forced to dedicate resources to continuing a precarious existence rather than move themselves out of poverty, contributing to psychological afflictions such as depression (Adelekan 2010).

As elsewhere, the recursive cycles of poverty exacerbated by climate change-induced water issues afford an excuse for the government to clear slums and sell the land to developers. Despite using its low-lying situation as a reason for destroying the Maroko slum in 1991, the land became a development site for a new, high-income residential district (Adelekan 2010). On top of this tendency, economic infrastructure and important business investments lie in Lagos' hazard-prone areas.

While the impacts of flooding unevenly impact the poor, women in particular bear the brunt of its effects. According to Ajibade and McBean (2014), women do not believe that they suffer more from flooding, but floods cause disproportionate disruptions to women's access to food and economic security while also suffering greater anxiety over childcare. For women, a decline in hygiene and privacy can compound other social risks, and maternal health can decline with sometimes life-threatening consequences (Ajibade et al. 2013).

According to Ajibade et al. (2015), the impacts of climate hazards such as flooding are felt not only on individuals but also on larger social formations. Analyzing behavioral factors such as "reasons for living in the area of residence, engagement in coping, adoption of coping strategies and participation in community development" and "influences of socio-economic factors (i.e. respondent's education, occupation and income) on flood impacts," the report found that "socio-cultural/demographic factors such as gender, ethnicity, religion and family structure are important interveners of the intensity of flood impacts" (Ajibade et al. 2015) Hence, the strength of social networks contributes to people's capacity to withstand climate hazards, and their absence can lead to mounting personal crises. Because flooding can uproot communities, the lack of durable social networks becomes a product of climate hazards.

## 7.7 Potential Responses

Planning for climate-induced water issues in Lagos is complicated by the fact that, after spreading inland northward and westward, the conurbation has become a megacity. The challenge for planners is recognized by the government, which advocates a philosophy grounded in political will and public participation with an increasing focus on climate change. Perhaps the turning point in Nigeria's climate awareness came in 2007, when their membership in the C40 Large Cities Climate Leadership network marked a movement to set theory into action (Mehrotra et al. 2009).

The concomitant formation of the Ministry of Environment's climate change unit, which organizes outreach for young people and organizes an annual summit (Elias and Omojola 2015). With an ecumenical approach to identifying hazards, expanding adaptive capacity, networking groups, promoting job growth, and educating, these summits provide a useful landmark for organizing efforts in Lagos that draw students, educators, political leaders, development partners, international investors, and members of the lay public.

Some contention exists as to how to proceed with assessing and prioritizing vulnerability to climate change. While some argue for a simple, hazard-based approach, which focuses on the exposure of people and infrastructure to climate threats, other scholars make the case for "a housing rights-based approach in evaluating vulnerability and adaptation to climate change in slum communities" (Ajibade and McBean 2014). Just as slums are built in hazardous areas due to restrictive land rules, the fact of their poverty renders them more vulnerable and furthers their exposure to climate change. Similarly, scholars insist that the response to these disproportionate impacts of climate change on women must include a human rights–based approach that reinforces wage equality, adequate social development, and more equitable distribution of household tasks and resources, as well as the "promotion of women's self-reliance" (Ajibade et al. 2013).

In addition, efforts by the state government, Lagos State Waste Management Authority, and the Nigerian Meteorological Agency to spread warnings about impending floods tend to lack effective dissemination (Onuminya and Nze 2017). The radio and television transmissions do not get through to the most vulnerable, who tend to lack electricity. Warnings adorning buses can miss their desired targets due to lack of road access. And many simply do not trust the government, so they ignore the messages outright. As well, the lack of adequate shelters in the event of inclement weather makes warnings less useful.

Based on these problems, scholars call for an integrated and holistic approach to slums, not only recognizing their role in the urban fabric, including them in urban development by upgrading infrastructure and houses to meet environmental challenges, but also improving the place of women in society. Insofar as international collaborations and organizations can help further these goals, the Nigerian government and Lagos State insist they are working to improve conditions by tackling complex, interconnected systems of social-ecological-technological vulnerability with integrated approaches to human rights and water infrastructure development.

## 7.8 Conclusions

Despite the promise of a workable approach to resilience, some contradictions embedded within the development of Lagos from the start continue to press against the boundaries of feasible reforms. Despite the numerous people living in desperate poverty and living in the intensification of climate hazards every year, Lagos is also becoming a magnet for the world's billionaires. In 2000, Lagos constructed a sand island known as "Banana Island," off the coast of the Ikoyi neighborhood, which looks back from mansions and penthouses worth more than a thousand US dollars per square meter (Fasona et al. 2020).

The few detached houses sold on the island go for US $2.75 million at the lowest, with the most expensive listings stretching into billions of dollars (Ruhling 2019). Because income is one of the determining factors in bio-psychosocial resilience to flooding in Lagos, the growing income gap openly flaunted by the development of an offshore island for the super-rich does not indicate that economic inequity is a serious concern for policymakers and developers in Lagos.

As in so many other places, it does appear that for Lagos to develop a clear and programmatic approach to climate resilience, equity must be on the table. Housing and women's rights must be foregrounded in any effort for sustainable urban development in order to counteract the major problems associated with flooding and sea-level rise, which have broader impacts on the entire state. Lagos faces many challenges, but through the furtherance of current efforts and the continued application of inclusive and participatory development philosophies, the megacity will emerge as a global leader in preparedness.

## References

Abiodun S (2021) Impacts of flood on inhabitants living around Lagos lagoon corridors: emphasis on makoko, ido, and oto. Journal homepage: www.ijrpr.com. ISSN, 2582, 7421

Adelekan IO (2010) Vulnerability of poor urban coastal communities to flooding in Lagos, Nigeria. Environ Urban 22(2):433–450

Adelekan IO (2016) Flood risk management in the coastal City of Lagos, Nigeria. J. Flood Risk Manage 9(3):255–264

Adewole AT (2009) Waste management towards sustainable development in Nigeria: a case study of Lagos state. Int NGO J 4(4):173–179

Agboola JI, Ndimele PE, Odunuga S, Akanni A, Kosemani B, Ahove MA (2016) Ecological health status of the Lagos wetland ecosystems: implications for coastal risk reduction. Estuar Coast Shelf Sci 183:73–81

Ajibade I, McBean G (2014) Climate extremes and housing rights: a political ecology of impacts, early warning and adaptation constraints in Lagos slum communities. Geoforum 55:76–86

Ajibade I, McBean G, Bezner-Kerr R (2013) Urban flooding in Lagos, Nigeria: patterns of vulnerability and resilience among women. Glob Environ Chang 23(6):1714–1725

Ajibade I, Armah FA, Kuuire VZ, Luginaah I, McBean G, Tenkorang EY (2015) Assessing the bio-psychosocial correlates of flood impacts in coastal areas of Lagos, Nigeria. J Environ Plan Manag 58(3):445–463

Ayeni AO (2014) Domestic water source, sanitation and high risk of bacteriological diseases in the Urban slum: case of cholera in Makoko, Lagos, Nigeria. J Environ Pollut Hum Heal 2:12–15

Bigon L (2005) Sanitation and street layout in early colonial Lagos: British and indigenous conceptions, 1851–1900. Plan Perspect 20(3):247–269

Bigon L (2008) Between local and colonial perceptions: the history of slum clearances in Lagos (Nigeria), 1924–1960. Afr Asian Stud 7(1):49–76

Bigon L (2009) Urban planning colonial doctrines and street naming in French Dakar and British Lagos c. 1850–1930 ABSTRACT. Urban Hist 36(3):426–448. https://doi.org/10.1017/S0963926809990125

Bollaert C, Aliyu T, Cascant-Sempere MJ (2023) Embedding research ethics into an international development programme: a case study of evidence and collaboration for inclusive development (ECID) in Nigeria. Community Dev J 58(1):121–135

Campus I (2009) Challenges of sustainable physical planning and development in Metropolitan Lagos. Journal of Sustain Dev Dept Urban Reg Plan Fac Environ Technol 2(1):160

Elias P, Omojola A (2015) Case study: the challenges of climate change for Lagos, Nigeria. Curr Opin Environ Sustain 13:74–78

Fashae OA, Onafeso OD (2011) Impact of climate change on sea level rise in Lagos, Nigeria. Int J Remote Sens 32(24):9811–9819

Fasona MJ, Ariori NA, Akintuyi AO (2020) The challenge of urban evolution and land management in developing countries: some lessons from the City of Lagos, Department of Geography, University of Lagos, Lagos

Gandy M (2005) Learning from Lagos. New Left Rev 33:37

Gandy M (2006) Planning, anti-planning and the infrastructure crisis facing metropolitan Lagos. Urban Stud 43(2):371–396

Girshick PBA, Thornton J (2001) Civil war in the kingdom of Benin, 1689–1721: continuity or political change? J Afr Hist 42(3):353–376

Hopkins AG (1966) The Lagos strike of 1897: an exploration in Nigerian labour history. Past Present 35:133–155

Ibem EO (2011) Public-private partnership (PPP) in housing provision in Lagos Megacity Region, Nigeria. Int J Hous Policy 11(2):133–154

Ikomi RB, Odum O, Erueseraise M (1997) Fish communities of the Ovwere stream in The Niger Delta area, Nigeria. Acta Ichthyol Piscat 27(2):113–125

Ikuemonisan FE, Ozebo VC, Olatinsu OB (2020) Geostatistical evaluation of spatial variability of land subsidence rates in Lagos, Nigeria. Geod Geodyn 11(5):316–327

Kolawole OD, Williams SB, Awujola AF (2010) Indigenous fish processing and preservation practices amongst women in Southwestern Nigeria. Indian J. Tradit. Knowl 9:668–672

Law R (1978) The career of Adele at Lagos and Badagry, c. 1807-c. 1837. J Hist Soc Nigeria 9(2):35–59

Manzoor J, Sharma M (2019) Impact of biomedical waste on environment and human health. Environ Claims J 31(4):311–334

Mehrotra S, Natenzon CE, Omojola A, Folorunsho R, Gilbride J, Rosenzweig C (2009, June) Framework for city climate risk assessment. In: Fifth urban research symposium, Marseille, France, pp 28–30

Obiefuna JN, Nwilo PC, Atagbaza AO, Okolie CJ (2013) Land cover dynamics associated with the spatial changes in the wetlands of Lagos/Lekki Lagoon system of Lagos, Nigeria. J Coast Res 29(3):671–679

Olukoju A (2017) Making sense of the Yoruba Littoral. Yoruba Stud Rev 2(1):45–60

Oluwasegun JM (2017) Managing epidemic: the British approach to 1918–1919 influenza in Lagos. J Asian Afr Stud 52(4):412–424

Omogunloye OG, Ayeni OO (2012) Geospatial analysis of the distribution of primary water pipelines in Lagos State (A case study of all local government areas in Lagos State). J Environ Sci Res Manage 4:87–110

Onuminya TO, Nze EC (2017) An appraisal of waste management in Lagos metropolis: a case study of Lagos state waste management authority (LAWMA). Age 15(24):71

Opeyemi RS (2017) The links between variations in climate patterns and ITCZ position over Nigeria. World Sci News 87:191–204

Oshodi L (2013) Flood management and governance structure in Lagos, Nigeria. Regions Mag 292(1):22–24

Oteri AU, Ayeni RA (2016) The Lagos megacity. Water, Megacities, and Global Change, 1–36

Ruhling N (2019) Banana Island in Lagos is a billionaire's paradise. Mansion Global, February 2, 2019

Smith W (1980) Master plan for metropolitan Lagos. United Nations Centre for Human Settlements (Habitat)

Tejuoso OJ (2006) Wetland uses/dynamics for agricultural purposes and its health implications in lower Ogun river basin, Lagos, Nigeria: a technical report

Twumasi Y, Merem E, Namwamba J, Mwakimi O, Ayala-Silva T, Abdollahi K, Okwemba R, Lukongo O, Akinrinwoye C, Tate J, LaCour-Conant K (2020) Degradation of urban green spaces in Lagos, Nigeria: evidence from satellite and demographic data. Adv Remote Sens 9:33–52. https://doi.org/10.4236/ars.2020.91003

Uwa J (2018) Transcultural tension and the politics of sewage management in (post) colonial Lagos. Soc Dyn 44(2):221–238

Van Tol D (2007) Mothers, babies, and the colonial state: the introduction of maternal and infant welfare services in Nigeria, 1925–1945. Spontaneous Gener A J Hist Philos Sci 1(1):110–110

Wasserman RJ, Dalu T (2022) Tropical freshwater wetlands: an introduction. In: Fundamentals of tropical freshwater wetlands. Elsevier, pp 1–22

Yusuf MA, Abiye TA, Butler MJ, Ibrahim KO (2018) Origin and residence time of shallow groundwater resources in Lagos coastal basin, south-west Nigeria: an isotopic approach. Heliyon 4(11):e00932

# Chapter 8
# Cape Town, South Africa

**Abstract** Cape Town maintains a large and economically successful metropolitan area supported by more than two centuries of hydrologic infrastructure. Yet, the impacts of climate change have already been felt in the disastrous drought of 2018 followed by the record-breaking heat of January 2022. This chapter discusses the climate hazards of floods and droughts experienced by Cape Town as a result of interannual climate teleconnections amid the expansion of the Hadley cell. The development of the state's water storage is traced, like many rapidly growing cities, to a staggered development precipitated by periods of flooding and drought. The worse the climate hazards become, the more important the storage, yet heat and dryness increasingly threaten the water supply, blunting the effect of increased reservoir capacity. As attention turns to water conservation mechanisms, the problem of scarcity exacerbates long-standing political conflicts while throwing the problem of inequality into stark relief. We conclude that the city can take greater steps to contend with flooding and droughts but must deploy resilient strategies in conditions of water scarcity to avoid disproportionate impacts of water policies on the poor.

**Keywords** Apartheid · Day zero · Drought · Hadley cell · Disaster management

## 8.1 Introduction

The City of Cape Town in South Africa grabbed headlines in 2018 as the municipality faced its worst drought-induced crisis in modern history (Maxmen 2018). The crisis brought to a head years of concern over climate-induced weather extremes, drawing attention to their causes and implications. Although the drought broke—so much so that dams were brimming with water by October 2020—the scale and scope of the disaster raised awareness as to the interlocking nature of human–water systems and their vulnerabilities.

Intermittent drought has always played a role in the piecemeal development of the Western Cape's hydrologic systems, as population growth conspired with low-flow years to necessitate the incremental construction of new reservoirs every few



H. Chang, A. R. Ross, *Climate Change, Urbanization, and Water Resources*, https://doi.org/10.1007/978-3-031-49631-8_8

decades around the City of Cape Town and ultimately expanding outward with more extensive tunnels and pipelines (Ibebuchi 2021; Grant 1991). Eventually, dam construction facilitated greater reservoirs, satisfying demand in the short term but ultimately falling short of demand.

The story of Cape Town's development involves the intermittent tensions between English, Boer, and native African groups and the staggered construction of hydrological infrastructure to meet the accelerating pace of growth. As Cape Town expanded, informal settlements developed in flood-prone areas, marking the hazards of pluvial flooding as a second major water issue in the city and adding the consequences of spatial inequality to those of economic and political inequality (Dube et al. 2022).

## 8.2 Climate and Physical Geography

Lying at the southwestern extreme of South Africa, Cape Town faces the Atlantic Ocean, extending about 154 square miles in all directions. Toward the south, the city relinquishes its hold on the land to Table Mountain, which juts out of the ocean in a hook-like peninsula that embays nearly 11 cubic miles. Called False Bay for its capacity to fool sailors into thinking it was part of the ocean (until they came upon Table Mountain's majestic facade), this water body lies at the westernmost edge of the shallow Agulhas Bank, which juts into the ocean about 150 miles before steeply descending into the ocean's abyss (Pfaff et al. 2019). It is a pivot point of the global climate system that involves four intersecting climatological and oceanic processes across differential spatial and temporal scales: the Agulhas current and its related eddies; the expansion of the Hadley cell; and the variability of the Southern Annular Mode.

The site is ideal not only for staging maritime expeditions into the Indian Ocean but also for fishing, due to an intense coastal upwelling of cold waters from the ocean's deep with both atmospheric and oceanic influences. On the one hand, the area's warm waters and low-pressure systems bring about cyclonic storms whose winds, channeled by the coast's mountains, shear the shallow waters away from the coastal shelf, causing deep waters to rise up and replace them (Reason 2001; Malan 2017). On the contrary, easterly eddies ripping away from the warmer Indian Ocean contribute to processes of mixing from the surface down (Arruda et al. 2014).

The eddies are caused by the prevailing current, which spills the warm, salty Indian Ocean into the colder Atlantic, causing the waters to retroflect back eastward just south of Cape Town (Biastoch and Krauss 1999). As a result of the retroflection, strong eddies are shed around the numerous promontories of the bank, bringing warm water westward into the Atlantic and making the waters particularly perilous for seafaring. It is not a coincidence that the "Cape of Good Hope" is thusly described, because the difficulties with eddies are only compounded by the forceful prevailing winds of the "Roaring Forties" to the south (Boebel et al. 2003).

These eddies that bring salty, warm water into the Atlantic also appear to help control the global conveyor belt known as the thermohaline circulation (or meridional overturning) (Schouten et al. 2002). A massive deep-ocean river, the thermohaline circulation, begins amid the formation of ice sheets in the Arctic, as the salt rejected during freezing causes the colder water to become denser and sink down, traveling meridionally north-south and then sweeping into the Indian Ocean and wrapping around the Hawaiian Islands of the Pacific before returning to the Atlantic. Since it brings warmer water back to the poles, the Agulhas Current helps govern the formation of sea ice, which in turn reflects solar energy and cools the world's climate system (Lutjeharms 2007).

Unfortunately, rapidly melting sea ice at the poles slows down the thermohaline circulation, which could have global impacts. However, the increased atmospheric circulation in the Indian Ocean caused by warmer climates and enhanced evaporation accelerates winds that drive the Agulhas current, causing the current to broaden, possibly offsetting the slowed thermohaline (Beal and Elipot 2016). At the same time, the Agulhas current's extra force causes steeper, more powerful waves in the already tempestuous seas. But the current's forcing can also reinforce the intense coastal upwelling, providing nutrients for plants needed to sustain higher fish habitat (Goschen et al. 2015). Along with the upwelling, however, the amplified Agulhas current also corresponds to sea surface temperature anomalies off the coast, which intensifies thunderstorms in nearby communities, such as Cape Town (Nakamura 2012).

Meanwhile, another devilish process works its way into the Cape. Normally, the warm tropics cause evaporated water vapor to rise with the air around the equator, wringing out from the poleward-venturing weather systems of the troposphere before the dry air finally descends at roughly the 30th parallel. These mid-latitude high-pressure zones of descending, dry air make the northern Sahara, Gobi, and Sonora Deserts—and their southern counterparts, the Great Sandy Desert, barren Atacama, and South Africa's own Kalahari Desert—so arid. The process of equatorial uplift and mid-latitude descent is collectively known as the "Hadley Cell." And it appears that with climate change, the Hadley cell may be expanding further poleward, potentially putting Cape Town in its sites (Pascale et al. 2020; Mahlobo et al. 2019).

As mid-latitude pressure ridges drift poleward, they may clash with existing inter-annual variability cycles associated with atmospheric circulation around the polar cap. This Antarctic oscillation, otherwise known as the Southern Annular Mode, involves the rapid flow of air around the South Pole in its positive phase and slower movement further from the Pole in its negative phase. Scientists find a correlation between the negative Annular Mode and rainy seasons in Cape Town, indicating that, with observations of El Niño-like variability and cyclonic weather systems from the Atlantic, Antarctic oscillation plays an important role in the position of the jet stream, ultimately setting the ebb and flow of South Africa's tempestuous floods and scorching droughts (Sousa et al. 2018).

Thus, Cape Town is pressed between contrasting forces—the disruption in weather patterns linked to the Agulhas current and the poleward expansion of the

Hadley cell appear to conspire with the Southern Annular Mode to intensify the sharpening of climatic variability. The increased frequency of late-winter high-pressure systems associated with dry weather that contribute to the risk of wildfires and smog-causing inversions may play the foil to severe flooding in certain years. Thus, with temperatures projected to increase by 3 °C, the metro area has experienced an intensification of floods and droughts along with an increase in demand for potable water (Perkins-Kirkpatrick and Gibson 2017).

## 8.3 Colonial History

Tumultuous seas and bitter weather have always characterized the relationship between the Cape and the Europeans who established themselves there. Agulhas gets its name from the Portuguese word for "needles," describing the numerous capes that jut out into the ocean and provide sharp obstacles for navigators attempting to avoid the fast 40th parallel. First arriving in the late fifteenth century, Portuguese explorers attempted to dominate the region but found themselves pushed back by the indigenous Khoekhoen peoples (Verbuyst 2016). The area remained a stopping point for expeditions from Europe to Asia, but it was not until 150 years later that the Dutch finally mustered the necessary resources to construct the earliest fortifications for a supply post (Abrahams 1993).

An ethnic group tied by their common Khoe language, the Khoekhoen lived in the area prior to the apparent Bantu migration that took place between 2000 and 3000 years ago. A diverse group, some Khoekhoen herded Nguni cattle, sheep, and goats, while others foraged shellfish and still others led relatively sedentary lifestyles (Badenhorst 2006). European travelers would trade tobacco and other goods for fresh meat from the Khoekhoen pastoralists, but after commerce increased and settlements expanded beginning in the 1650s, native populations declined due to the spread of smallpox (Webley 1992).

Because the Cape receives light rain in the summer, settlers found its rivers low during the dry season, and drought in 1663 proved destabilizing. To compensate, the local commander Zacharias Wagenaer ordered the construction of a reservoir on Table Mountain some 130 feet by 45 feet by Khoekhoen workers and other laborers. By 1670, a channel made of stone took water from the reservoir to the docks, where sailors could easily obtain the resource without carrying barrels. Remnants of the old system can still be seen in Cape Town's city center (Fig. 8.1).

When not in drought years, settlers relied on the Liesbeek River for most of their agriculture, drawing also from the Sandvlei's tributaries and the Disa River around the Hout River system (Bottaro 1996). Thus, the area opened to incoming migrants from all over the world. As migration from Asia increased and Huguenots arrived fleeing repression under the French crown, settlers enclosed new territories for a burgeoning wine industry as well as food production, and the Dutch East India Company brought slaves in from Java and Madagascar to labor in the fields (Fourie and Fintel 2011). Amid the tenuous and shifting alliances and conflicts between

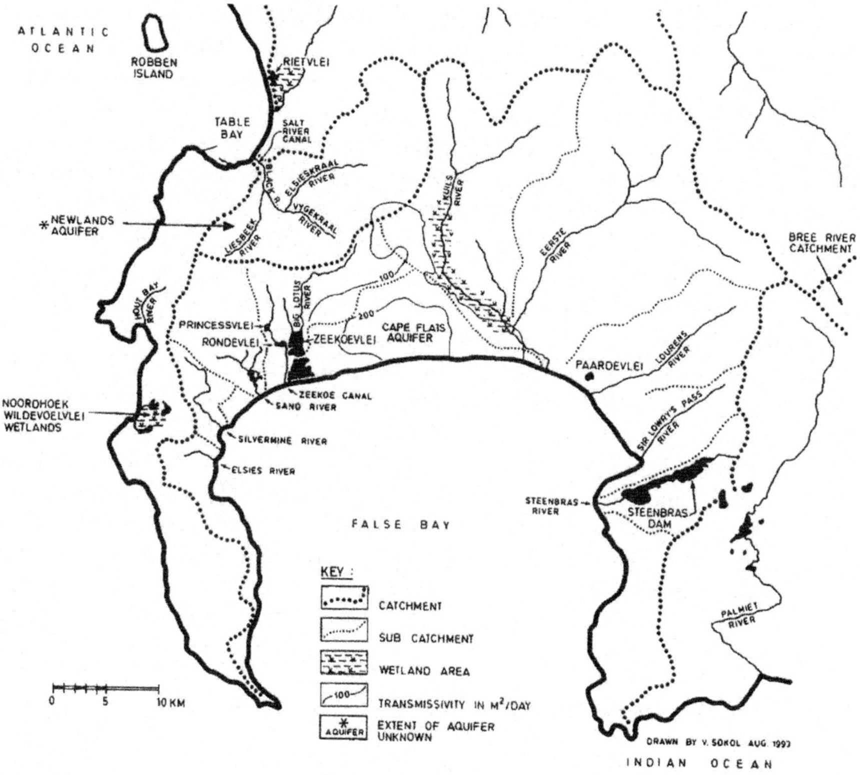


**Fig. 8.1** Water resources, water infrastructure, wetland, and catchment boundaries in Cape Town. (Source: Quick 1995)

European powers that marked the late eighteenth century, the Dutch invited the English in to defend against the French. The English subsequently invaded and overtook the Cape, and, after handing it back for a brief period, invaded it again and took power over the imperial property in order to guarantee their interests in India.

Under British rule, the Cape Colony expanded through wars against the indigenous Africans, constructing infrastructural developments that facilitated the growth and development of Cape Town as the primary center of the region. In 1814, the fetid nature of some low-flow riparian systems used for sewage brought the local government to install a reservoir and iron pipe system in the city, complete with fountains that residents could use to collect potable water, rather than traveling to polluted rivers (Brodie 2015). Within 20 years, however, water shortages emerged. The fountains were covered by pumps, and by 1850, the city halted street service to ensure sufficient water for fires. New city planners constructed two larger reservoirs holding 13 million gallons. However, the city's growth outpaced hydrologic production, leading to a kind of infrastructure race to keep up with demand (Juuti et al. 2007).

Despite such efforts to expand water access, British rule roiled the local Dutch population, known as Boers, who rejected free trade and "self-taxation" policies and revolted against the enforcement of protections for servants against exploitation and cruelty. It was Great Britain's emancipation of slaves across the colonies in 1834, however, that prompted the "Great Trek"—a mass migration of Dutch-descended people out of the Cape Colony and into the Transvaal. With the discovery of gold and diamonds in the Transvaal, the Boer states grew wealthier, particularly Johannesburg, but the Cape continued to encroach onto lands the Boer had settled. Two sanguinary Boer Wars ensued, lasting from 1880 to 1881 and 1899 to 1902, ultimately resulting in a British victory and control over the former South African Republic and Orange Free State (Fremont-Barnes 2014).

The challenge to Johannesburg and Boer hegemony ensured Cape Town's position as a power center in South Africa. However, water shortage crept into the life of the city and its growing suburbs by 1905, leading to new schemes for water storage. In 1908, the Kloof Nek reservoir was created and the Cape's first dam, called the Steenbras Dam, was built from 1918 to 1921 (Brodie 2015). The Steenbras Dam became the city's leading water supply but was raised in 1928 to continue to keep up with demand, despite ongoing and severe differentiation between European and native areas (Enqvist and Ziervogel 2019).

## 8.4 Modern Infrastructure and Climate Demands

Reconciliation efforts between English and Boer enfranchised a degree of sovereignty for Boer political systems, which mobilized the legacy of the war and its military figures to build an electoral base. Heavy handed in some ways and weak willed in others, British command over the region deteriorated apace. To continue building up Cape Town, the local government began to construct hydrologic enterprises that ensured water storage and distribution for the growing population. Yet, this expansion of water infrastructure began to wane in the 1990s, even as economic migrants flocked into the city, causing a perfect storm and became known as "Day Zero."

Following three consecutive dry years in the 1930s, planners seeking to secure water resources for the growing city drafted plans for a new dam at the Wemmershoek catchment in the Drakenstein mountains to the east of Cape Town. A prospective dam site for over 50 years, the Wemmershoek seemed like the most opportune place until a flood damaged construction and set back the timetable—a critical problem during the race to avert a water crisis. By 1957, however, the dam was complete and remains standing today (Van Vuuren 2010).

Within 15 years, however, Cape Town had exceeded 1.1 million inhabitants, and another three consecutive dry years during the early part of the decade caused water shortages, leading to water restrictions during summer months. New dams were set forward in Voelvlei and Theewaterskloof in 1971 and 1978, respectively, with the latter serving as the largest hydrologic structure in the water system. In another

20 years, three consecutive dry years reoccurred, followed in turn by plans for a new dam in the Berg River, which was delayed until 2009 due to trends in environmental priorities. While models analyzing patterns of population growth and water consumption called for an expansion of the city's water infrastructure by 2015, local authorities increasingly shifted toward water conservation rather than dam construction to ensure sustainability. The crisis was clear: Between 1996 and 2017, the city's population increased by 67% despite adding only 15% to the water supply over that period (Nhamo and Agyepong 2019).

In the 1990s, the number of undocumented migrants to South Africa increased. A concomitant xenophobic reaction emerged among politicians, even in places such as Cape Town that are usually known for their cosmopolitanism. The failure to expand water infrastructure in tandem with the growing population thus resulted in part from the failure to accept the reality of an urban expansion and worsened the tendency to blame immigrants for the stressors of development. Also, the deeply entrenched legacy of *apartheid* that carried forward economic inequalities between South Africans of European and African descent included inequities in the water system. For instance, poorer schools are equipped with worse water infrastructure and are less maintained, using more water as a result (Booysen et al. 2019).

And in 2015, the all too predictable, though evermore intense, dry period set in, staying for three years. Rainfall near the two major dams plummeted from 1250 mm to just 500 mm between 2013 and 2015. The next year, it rose to 750 mm but fell back to 700 mm in 2017. By the end of 2017, streamflow in some catchments had declined to 20% of its normal rates. The city placed a ban on nonessential water use and hiked water tariffs by 26% (Muller 2017). To stave off "Day Zero," at which point the taps would shut off, local authorities requested residents keep consumption below 450 ml per day. Consumption remained above 500 ML, and despite the city's efforts to spread awareness of water restrictions throughout Cape Town, by 2018, only 55% of citizens actually followed the city's water mandates (Nhamo and Agyepong 2019).

Despite some recalcitrance, however, the city, itself, brought water usage down from 1.2 billion liters per day to an incredible 511 million liters per day. Along with residential restrictions, Cape Town was supported by augmentation programs that reused treated wastewater, while also stopping water use in three irrigation districts. In addition, the city issued water management devices in one out of five houses in Cape Town, cutting their water flow after a monthly limit was reached (Robins 2019).

In the meantime, a water donation program began, as other areas around South Africa began sending bottled water to relieve the plight of Cape Town residents. As Day Zero loomed nearer, the city sent police to collection points distributed throughout the city to stave off potential "anarchy," even as some activists denied the reality of Day Zero entirely. To these anti-privatization activists, Day Zero simply existed to threaten the public into accepting the installation of water-saving measures and ultimately to sell off the city's water supply and management to corporations (Robins 2019).

While some of their claims may have been excessive, the activists hit on crucial inequalities within South Africa's water system. Formal settlements account for

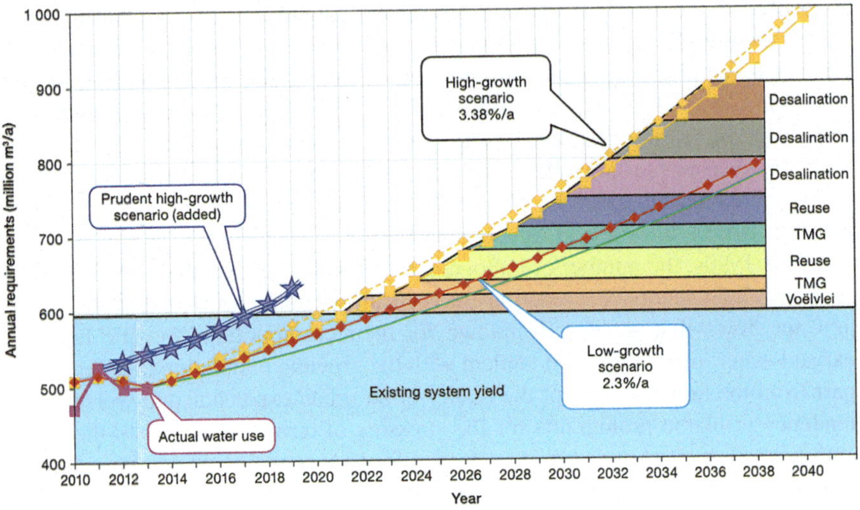


**Fig. 8.2** Projected water demand under different growing scenarios in Cape Town. (Source: Muller 2017)

more than 60% of Cape Town's water consumption, with informal settlements housing the very poor using less than 5% of the city's water. Yet, the poor came under the same rigorous scrutiny and water restrictions as did the relatively well-to-do, and low-income women bore the brunt of the deprivation. Women working in domestic labor, for instance, now washed clothes by hand, as the general tasks of cleaning, washing, and watering grew increasingly difficult. And while the wealthy could afford to drill boreholes and supplement municipal water cutbacks, such conveniences did not exist for the poor and marginalized. Since 39% of Cape Town lives in poverty, inequities within the city's water response were as widely felt as they were austere (Wallace 2021) (Fig. 8.2).

## 8.5 The Specter of Floods and Adaptation

As mentioned earlier, climate variability in the Western Cape leads to rainy periods in Cape Town along with periods of extended drought. Like drought, the heaviest impacts of flooding falls on the poor, many of whom live in the Cape Flats, an area with a high water table and little drainage infrastructure (Drivdal 2016). Most of Cape Town maintains adequate storm drains and planning to prevent nuisance floods, whereas, in Cape Flats, nuisance floods can lead to costly damage and even death (Ziervogel 2019). To meet the demands of such climate-induced impacts, South Africa has adopted a decentralized approach to adaptive management regarding both floods and droughts. While these approaches have succeeded in some efforts, their ambitious multi-stakeholder efforts can pose important challenges.

After the fall of the apartheid regime in 1994, South Africa implemented upgrades to informal settlements to enfranchise housing as a human right. In 2008, Cape Town counted more than 100,000 people in 300 informal settlements, most of which lay in the Cape Flats (Drivdal 2016). In addition to the geomorphic reasons for flooding during heavy rains, infrastructure built in informal settlements also tends to clog, causing nuisance flooding during even moderate rain (Musungu et al. 2016). Since the centralized approach to improving infrastructure in areas most impacted by flooding can contribute to added complications, engaging with multiple civil society and governmental groups can provide crucial openings, as well as difficulties (Goncalo et al. 2007).

Diverse networks that collaborate across the ordinary boundaries between civil society and government can provide the flexibility needed to ensure that planning and implementation of resilience frameworks can take root in society, developing the necessary learning and practices throughout the public sector to mitigate the worst impacts of climate change. However, shortfalls in staffing, formal governance structures, cultural norms, public participation, and communication can impede network structures. By contrast, access to financial, human, and social capital, including money, awareness, and access to social networks, promote adaptive measures in a decentralized fashion (Desportes et al. 2016).

Unfortunately, political contestation within informal settlements such as Sweet Home can create barriers to resilience. When the African National Congress, for instance, holds hegemony within local councils, they can distribute emergency resources as political favors (Desportes et al. 2016). In lieu of more corrupt motives, local leaders also face immense challenges and pressures from overlapping political and social commitments relative to finite resources, which city officials often fail to recognize (Drivdal 2016). Also, the City of Cape Town created canals within settlements to make clearing blockages easier, believing that locals would otherwise dispose of solid waste hazardously, but locals dislike the canals due to their stench and health issues (Desportes et al. 2016). In some cases, residents simply prefer relocation to upgrades made possible by channeling money through NGOs and community groups (Musungu et al. 2016). When the interests of the city and those of the residents do not align, complications occur that can lead to bitterness and distrust, which cause multi-stakeholder systems to break down (Desportes et al. 2016).

Just as with the Day Zero crisis, distrust leads to impasses, as efforts to expand social learning and increase adaptive capacity falter. In this situation, a multi-stakeholder effort can actually enhance fragmentation and conflict (Drivdal 2016). While the solution that would cut the Gordian knot would most likely be relocation, Cape Town's lands are all subject to competition from other claims involving different forms of land use (Ziervogel 2019). Political will, then, lies more with short-term in situ adaptations than with meaningful, long-term goals that would require broader forms of structural adaptation.

## 8.6 Different Problems, Complex Solutions

Water shortage and flooding pose dual problems likely to intensify under climate change in South Africa. While cyclical periods of drought may worsen as the mid-latitude arid band expands poleward, storms related to the Southern Annular Mode may increase flooding issues in the Cape Flats. While floods and droughts pose two separate problems—particularly in terms of structural relations, given the dependence of flood regimes on informal settlements—their solutions may overlap in the expansion of multi-stakeholder efforts to increase social learning and strengthen community engagement.

South Africa's 2015 Disaster Management Amendment Act defines climate change as, "[A] change in the state of the climate that can be identified by changes in the variability of its properties and that persists for an extended period, typically decades or longer." Adaptation to climate change is thus described as the "process of adjustment to actual or expected climate and its effects, to moderate harm or exploit beneficial opportunities." Adaptation to climate change, then, involves "the process of adjustment to actual climate and its effects" (Republic of South Africa 2015).

South Africa's disaster management agenda names groups such as the National Disaster Management Advisory Forum and South African National Platform for Disaster Risk Reduction, as well as the Defense Forces and police in assisting responses to climate-induced disasters. It also acknowledges the importance of traditional leaders in providing support for such crises (Nhamo and Agyepong 2019). For the City of Cape Town, disaster response to water-related disasters involves eight key points: minimizing the impact on life, dignity, and property; maintaining critical services; mitigating escalation of the crisis; reducing day-to-day effects; infrastructure protection; securing access to water for drinking and hygiene; limiting water-borne diseases; and prioritizing vulnerable residents (Nhamo and Agyepong 2019). Thus, Cape Town attempts to engage community representatives with a variety of governmental agencies and NGOs to adapt to water-related disasters by prioritizing services to vulnerable residents (Makaya et al. 2020).

At the same time, droughts and floods must be approached in different ways with similar strategic applications of infrastructure improvements and social learning. For drought mitigation, social learning becomes integral to disaster management. Since centralized efforts to restrict residents' water consumption tend to aggravate political dissent, efforts to educate people on the importance of independent self-regulation can help empower local communities to proactively adapt to hazards. In this regard, traditional leaders can help, but equitable strategies that do not unduly impact the poor must also be prioritized to avoid dissent. Still, structural improvements to water storage may also improve adaptive capacity for water shortages.

Infrastructure improvements can also help mitigate floods. In particular, adequate sewer infrastructure and drainage systems could be installed. However, social learning can also be developed to encourage residents of areas with a high water table to raise their domiciles and install flood prevention measures to protect their properties. On the other hand, local communities' political competitions and the complexity of nested political hierarchies must be considered to maximize outreach efforts while empowering appropriate leaders.

## 8.7 Conclusions

Poised in a unique climatological position, Cape Town faces a number of complicated challenges. The Agulhas current may be strengthening, causing increasing eddies to balance out the impacts of Arctic glacial retreat on thermohaline circulation. At the same time, the Southern Annular Mode may be intensifying, which, coupled with the expansion of the mid-latitude high-pressure zone, could spell chaotic conditions in the future.

In spite of the uncertainty and intensification, some patterns appear predictable—namely recurring droughts and pluvial flooding will persist and likely worsen, necessitating responsive infrastructure. The specter of continued climate-induced disaster poses a threat to the South African economy, as well as the security and health of its residents. Thankfully, administrators, water managers, politicians, and community leaders take these concerns seriously.

Yet, Cape Town is not the only place suffering the consequences of climate change in South Africa. Two years after the terrifying drought abated in 2019, Cape Town's dams maintained a fairly consistent level at around 70%, but on the other side of the country, in the Eastern Cape, water levels fell to historic lows. As Cape Town recovered, their neighbors slumped into their own disastrous impending Day Zero. The see-saw variability of droughts and floods may become the new normal, forcing the country to contend not only with local agencies, NGOs, and community leaders, but an even more integrated, multi-scalar network of federal, local, watershed, and municipal actors who can share resources, knowledge, and skills.

## References

Abrahams G (1993) The grand parade, Cape Town: archaeological excavations of the seventeenth century Fort de Goede Hoop. South Afr Archaeol Bull 48:3–15

Arruda W, Zharkov V, Deremble B, Nof D, Chassignet E (2014) A new model of current retroflection applied to the westward protrusion of the Agulhas current. J Phys Oceanogr 44(12):3118–3138

Badenhorst S (2006) Goats (Capra hircus), the Khoekhoen and pastoralism: current evidence from southern Africa. Afr Archaeol Rev 23:45–53

Beal LM, Elipot S (2016) Broadening not strengthening of the Agulhas Current since the early 1990s. Nature 540(7634):570–573

Biastoch A, Krauss W (1999) The role of mesoscale eddies in the source regions of the Agulhas Current. J Phys Oceanogr 29(9):2303–2317

Boebel O, Lutjeharms J, Schmid C, Zenk W, Rossby T, Barron C (2003) The Cape Cauldron: a regime of turbulent inter-ocean exchange. Deep-Sea Res II Top Stud Oceanogr 50(1):57–86

Booysen MJ, Wijesiri B, Ripunda C, Goonetilleke A (2019) Fees and governance: towards sustainability in water resources management at schools in post-apartheid South Africa. Sustain Cities Soc 51:101694
Bottaro J (1996) The changing landscape of the Liesbeek River valley: an investigation of the use of an environmental history approach in historical research and in classroom practice. Master's thesis, University of Cape Town
Brodie N (2015) The Cape Town book: a guide to the city's history, people and places. Penguin Random House South Africa, Cape Town
Desportes I, Hordijk M, Waddell J (2016) Improving flood risk governance through multi-stakeholder collaboration: a case study of Sweet Home informal settlement, Cape Town. South Afr Geogr J (Suid-Afrikaanse Geografiese Tydskrif) 98(1):61–83
Drivdal, L. (2016, September). Community leadership in urban informal neighbourhoods: micro-politics and micro-administration in informal settlements in Cape Town. In Urban forum (27, 3, 275-295). Dordrecht: Springer
Dube K, Nhamo G, Chikodzi D (2022) Flooding trends and their impacts on coastal communities of Western Cape Province, South Africa. GeoJournal 87(Suppl 4):453–468
Enqvist JP, Ziervogel G (2019) Water governance and justice in Cape Town: an overview. Wiley Interdiscip Rev Water 6(4):e1354
Fourie J, Fintel DV (2011) An unequal harvest: the French Huguenots and Early Cape wine-making. Tydskrif vir Geesteswetenskappe 51(3):332–353
Fremont-Barnes G (2014) The Boer War 1899–1902. Bloomsbury Publishing, London
Goncalo A, Bouchard B, Susienka M, Wilson K (2007) Improving flood risk management in informal settlements of Cape Town: an interactive qualifying project (Unpublished B.Sc. Project). Worcester Polytechnic Institute, Worcester
Goschen WS, Bornman TG, Deyzel SHP, Schumann EH (2015) Coastal upwelling on the far eastern Agulhas Bank associated with large meanders in the Agulhas Current. Cont Shelf Res 101:34–46
Grant D (1991) The politics of water supply: the history of Cape Town's water supply 1840–1920. Master's thesis, University of Cape Town
Ibebuchi CC (2021) On the relationship between circulation patterns, the southern annular mode, and rainfall variability in Western Cape. Atmos 12(6):753
Juuti PS, Mäki HR, Wall K (2007) Water supply in the Cape settlement from the mid-17th to the mid-19th centuries. In: Environmental History of Water-Global views on community water supply and sanitation. IWA Publishing, London, pp 165–172
Lutjeharms JRE (2007) Three decades of research on the greater Agulhas Current. Ocean Sci 3(1):129–147
Mahlobo DD, Ndarana T, Grab S, Engelbrecht F (2019) Integrated climatology and trends in the subtropical Hadley cell, sunshine duration and cloud cover over South Africa. Int J Climatol 39(4):1805–1821
Makaya E, Rohse M, Day R, Vogel C, Mehta L, McEwen L, Rangecroft S, Van Loon AF (2020) Water governance challenges in rural South Africa: exploring institutional coordination in drought management. Water Policy 22(4):519–540
Malan NC (2017) The impact of agulhas current dynamics on shelf waters: a modelling approach. University of Cape Town
Maxmen A (2018) As Cape Town water crisis deepens, scientists prepare for 'Day Zero'. Nature 554(7690):13–15
Muller M (2017) Understanding the origins of Cape Town's water crisis. Civil Eng (Siviele Ingenieurswese) 2017(5):11–16
Musungu K, Drivdal L, Smit J (2016) Collecting flooding and vulnerability information in informal settlements: the governance of knowledge production. South Afr Geogr J (Suid-Afrikaanse Geografiese Tydskrif) 98(1):84–103
Nakamura M (2012) Impacts of SST anomalies in the Agulhas current system on the regional climate variability. J Clim 25(4):1213–1229

Nhamo G, Agyepong AO (2019) Climate change adaptation and local government: institutional complexities surrounding Cape Town's Day Zero. Jàmbá J Disast Risk Stud 11(3):1–9
Pascale S, Kapnick SB, Delworth TL, Cooke WF (2020) Increasing risk of another Cape Town "Day Zero" drought in the 21st century. Proc Natl Acad Sci 117(47):29495–29503
Perkins-Kirkpatrick SE, Gibson PB (2017) Changes in regional heatwave characteristics as a function of increasing global temperature. Sci Rep 7(1):1–12
Pfaff MC, Logston RC, Raemaekers SJ, Hermes JC, Blamey LK, Cawthra HC et al (2019) A synthesis of three decades of socio-ecological change in False Bay, South Africa: setting the scene for multidisciplinary research and management, vol 7. Science of the Anthropocene, Elementa, p 7
Quick AJR (1995) Issues facing water resource managers and scientists in a rapidly growing coastal city: Cape Town, South Africa. S Afr J Sci 91(4):175–183
Reason CJC (2001) Evidence for the influence of the Agulhas current on regional atmospheric circulation patterns. J Clim 14(12):2769–2778
Republic of South Africa (2015) Disaster management amendment act 2015, Act No. 16 of 2015, Parliament of South Africa, Cape Town
Robins S (2019) 'Day Zero', hydraulic citizenship and the defence of the commons in Cape Town: a case study of the politics of water and its infrastructures (2017–2018). J South Afr Stud, 45(1), 5–29
Schouten MW, De Ruijter WP, Van Leeuwen PJ (2002) Upstream control of Agulhas Ring shedding. J Geophys Res Oceans 107(C8):23–21
Sousa PM, Blamey RC, Reason CJ, Ramos AM, Trigo RM (2018) The 'Day Zero'Cape Town drought and the poleward migration of moisture corridors. Environ Res Lett 13(12):124025
Van Vuuren L (2010) Wemmershoek-75 years in the making. Water Wheel 9(2):18–22
Verbuyst R (2016) Claiming Cape Town: towards a symbolic interpretation of Khoisan activism and land claims. Anthropol South Afr 39(2):83–96
Wallace B (2021) Avoiding day zero: how Cape Town cut its water usage by 50% in three years
Webley LE (1992) The history and archaeology of pastoralist and hunter-gatherer settlement in the north-western Cape, South Africa. Dissertation, University of Cape Town
Ziervogel G (2019) Unpacking the Cape Town drought: lessons learned. Cities support programmel Climate resilience paper. African Centre for Cities. University of Cape Town

# Chapter 9
# Melbourne, Australia

**Abstract** A city of more than 5 million people, Melbourne stands out as a global leader in water management, both on a theoretical and practical basis, facing multiple hazards from climate change. As such, it provides multiple lessons for the future of climate policy and water management based on its experiences, both positive and negative. Melbourne's position in southeastern Australia renders it vulnerable to El Niño-Southern Oscillation's (ENSO) interannual variability, as well as teleconnections with the Indian Ocean dipole. Together, these complex phenomena can reinforce or complicate patterns of dry and wet periods, leading to climate extremes shown to be exacerbated by climate change. In response, Melbourne took the lead in decentralized flood management through the development of "water-sensitive cities," including bioswales, mini-wetlands, and rain gardens. While some of these efforts have worked, studies also show that their piecemeal implementation and dependence on localized activism tends to lead to unequal distribution according to the interest and ability of communities to focus on such paradigms. Still, financial irregularities and political disputes make a concerted water management platform difficult that breaks from the historical trajectory of reactive infrastructure development.

**Keywords** Water-sensitive cities · Water-sensitive urban design · ENSO · Climate extremes · Water cycle

## 9.1 Introduction

In October 2022, massive flooding rocked the suburbs of the City of Melbourne, leading officials to evacuate thousands of people from their homes. Aided by a strong and prolonged La Niña system, the storm brought more rain in one day than typically seen over the course of a month (Jose and Jackson 2022). As river levels topped records, rescuers carried out hundreds of emergency rescues around the area. Moving from one crisis to the next, local administrators considered reopening COVID-19 quarantine centers for flood victims seeking shelter. In the aftermath, 70



H. Chang, A. R. Ross, *Climate Change, Urbanization, and Water Resources*, https://doi.org/10.1007/978-3-031-49631-8_9

people were displaced in the Maribyrnong suburb, including a refugee from World War II who told the state inquiry that the experience left him with post-traumatic stress disorder, reminding him of fleeing Nazi Germany (Ore 2023).

Climate change unquestionably poses a global threat, displacing huge numbers of people throughout the world, so that no one is safe from its onslaught. The case of Melbourne's rising climate extremes indicates the fearsome path of incoming climate change. Yet, Melbourne's efforts to combat the growing intensity of climate change–induced storms and droughts offer a slim beacon of hope to the rest of the world. While the efforts to secure "water-sensitive cities" have extended from Melbourne throughout the world, the large city of 5 million residents continues to struggle through political and economic snags on its way to promoting an equitable and adaptive response to the water issues portended by climate change.

## 9.2 Geography

Melbourne lies on the southeastern coast of Australia in the province of Victoria, wrapping around the Bay of Port Phillip. Some 30,000 years ago, the bay was dry due to lower sea levels, which also allowed indigenous peoples to traverse the land bridge to Tasmania and back, but after the recession of the Ice Age, the islands separated from the mainland, and the water returned to the bay. The temperate oceanic climate offers cool sea breezes and mild summers, and the Yarra River passing through the area helps irrigate native riparian ecosystems in tandem with coastal wetlands.

The Yarra River winds through about 32 miles of valley before emptying into Hobsons Bay at the northernmost point of Port Phillip Bay at the southern boundary of Melbourne. In the foothills east of Melbourne, Leadbeater's Possum, Southern Brown Bandicoot, Diamond Firetail, and other threatened, vulnerable, and endangered species thrive among 53 threatened bird species and numerous others amid a temperate rainforest ecosystem that ultimately rises up to peaks more than 1,400 feet high at Lake Mountain and a majestic 4,000 feet above sea level at Mount Donna Buang (SWIFFT 2021). Part of the Victorian Alps, which compose the southern portion of the Great Dividing Range, the Yarra Ranges developed more than 100 million years ago, by magma flows during the breaking apart of Gondwana (Slattery 2015).

The accessible bay and river made the site an opportune site for a settlement by colonists associated with the New South Wales penal colony. Dispossessing the aboriginal inhabitants, the new town grew through the proliferation of squatters. By the discovery of gold in 1851, the Port Phillip Region had separated from New South Wales, creating the colony of Victoria in the name of the Queen and issuing Melbourne the rights of the capitol. The gold rush brought an influx of settlers and an economic boom that persisted to the end of the century, bringing forth the Melbourne Hydraulic Power Company and, with it, pressurized, piped water (Pierce 2009). The pump houses that generated the pressure for hydraulics that could be used to power elevators in Melbourne's larger-than-life *fin-de-siècle* high rises marked only the fourth public hydraulic power system of its kind and would stay in operation until the 1960s (Gibson and Pierce 2009).

## 9.3 Super El Niño and Melbourne

In 1899, the seasonal monsoons failed over the Indian Ocean, causing widespread famine that devastated the subcontinent. Becoming director-general of observatories in India five years later, mathematician Sir Gilbert Walker endeavored to investigate the scientific causes of this dramatic climatological event using a wide-ranging array of weather data. What Walker found was an astonishingly broad causal pattern with its roots not in the Indian Ocean but in the tropical Pacific. With its most significant study points in Walker, Australia, and Tahiti, Walker's study would reveal among the most important causes of climate variability in the world—a system that would gain the name "Walker Circulation" after the scientist (Walker and Bliss 1932; Katz 2002).

The Walker Circulation depends on the trade winds that blow westward from Latin America. These winds push the tropical waters toward Southeast Asia, bringing the warm surface water with them and pulling cooler water from the depths of the Ocean up to coastal South America. As this cooler upwelling brings nutrients for plant life and an ideal ecosystem for fish near Peru and Ecuador, a pool of warm water accumulates in the Southwest Pacific, causing increased evaporation, humidity, and of course, storms. After the warm, saturated air rises into a low-pressure system over the warm pool, precipitation occurs, and the dried air diverges at the top of the atmosphere, ultimately descending over the central Pacific, creating high-pressure systems with calm, dry air.

But Walker's observations pointed to the later discovery that sometimes the trade winds break down, disrupting the circulation. Without the conveyor belt of winds, warm water sits in the eastern tropical Pacific. The swelling sea surface rejects any upwelling from the deep, and sea life subsides in favor of colder, nutritious waters elsewhere. The warm pool in the western Pacific breaks up, and evaporation decreases. The warm and wet low-pressure zone shifts to the central Pacific, sending tropical storms eastward to the South American coast, thus earning the turbulent title, "El Niño." Meanwhile, a high-pressure zone brings descending, dry air to the western Pacific. Instead of storms and floods, droughts descend, causing great damage with wide-ranging impacts in Australia. This sort of see-saw activity is called the "Southern Oscillation" (Neelin and Latif 1998).

While the intensity of El Niño events does not always match the intensity of ensuing droughts, it is clear that this "negative" phase of the Southern Oscillation caused by the breakdown in westward trade winds corresponds to hotter, drier weather for Australia. Located on the southern coast of the continent, Melbourne faces hotter temperatures as a result compounding its climatological challenges. At the same time, the decrease in cloud cover can cause a decline in back radiation, through which clouds absorb heat and emit it back to the earth's surface, thus causing colder nights. Also, wet season–inducing monsoons arrive weeks later, causing drier periods. Especially for southeast Australia, the increased heat and decreased storms lead to enhanced wildfires with major health risks for residents of Melbourne (BoM 2021; Mariani et al. 2016) (Figs. 9.1 and 9.2).

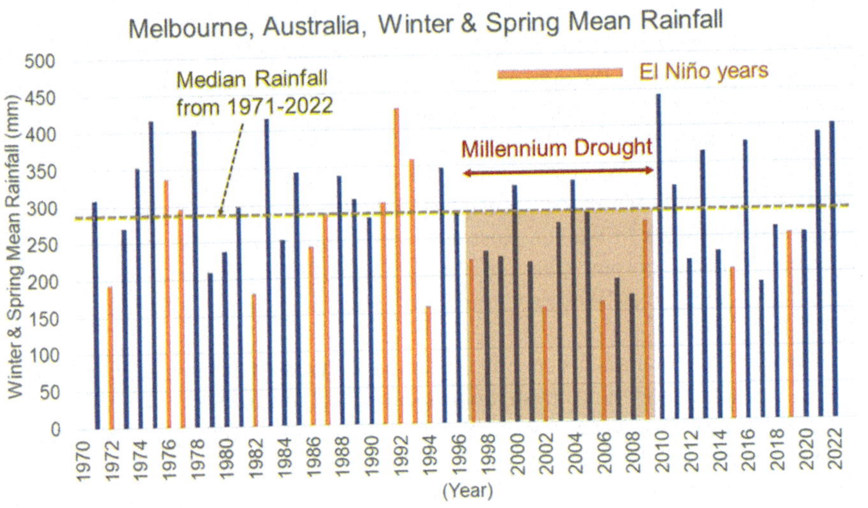


**Fig. 9.1** Melbourne's winter–spring mean rainfall in relation to El Niño events, 1971–2022. (Data source: BoM 2023)

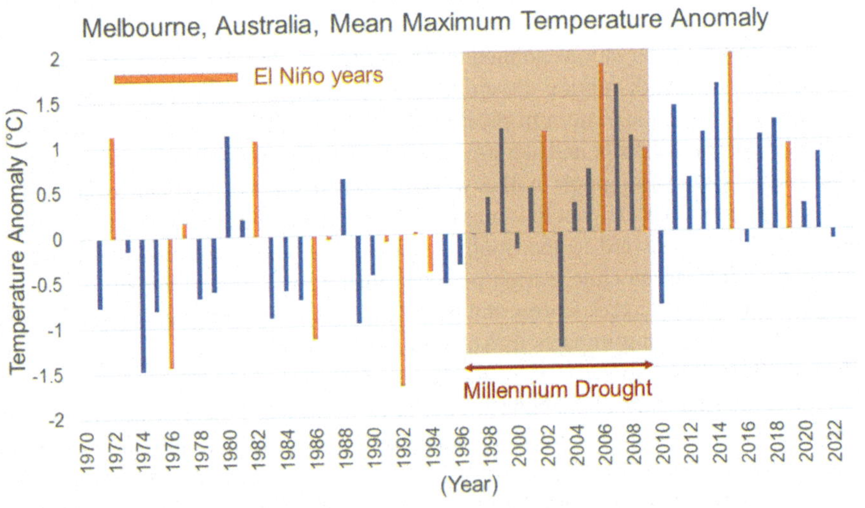


**Fig. 9.2** Melbourne's winter–spring mean maximum temperature in relation to El Niño events, 1971–2022. (Data source: BoM 2023)

The impacts of El Niño systems can be exacerbated by a corresponding positive phase of a similar variability system called the Indian Ocean Dipole (IOD). As the warm pool disperses from the western Pacific and air descends to create a high-pressure system, easterlies drive a warm pool across the Indian Ocean toward Eastern Africa, causing cyclones and flooding on the Horn. At the same time, drought conditions intensify in Australia, leading to worsened fire conditions. While the Indian Ocean Dipole is similar to El Niño, the two do not always occur simultaneously, but studies indicate that the IOD can trigger El Niño conditions across the Pacific and can contribute to "Super El Niño" systems (Hameed et al., 2018; Hong et al., 2014). This complex braiding of variabilities can cause unpredictable fluctuations in weather patterns, as well as headaches for climatologists.

## 9.4 Melbourne's Water

Unlike other areas covered in this book, such as Jakarta and Lagos, Melbourne will likely see not only enhanced floods or droughts but also an intensification of both due to climate change. The temperatures are projected to increase between 0.6 and 2.5°C by 2050. In addition, average annual precipitation will likely decrease by up to 13% by 2050 (Howe 2007). At the same time, the intensity of rainfall during storms will increase, along with the extreme of warm days. These changes will have clear implications for water resources in Melbourne.

Melbourne currently draws 70% of its water from the Yarra Ranges, known for its towering Mountain Ash trees. The Yarra Valley was originally called Birrarung by the Wurundjeri people, but explorers mistook the name of a specific falls to mean the name of the entire river, whereas Yarra-Yarra means something like "ever flowing," Birrarung approximately means "river of mist" (Briggs 2019). In what seems like an analogy for the larger settlement of the region, the Yarra falls would be ultimately dynamited by settlers seeking to harness the power of the river for transportation and dams (Sanders et al. 2021). The Thompson River, which stems from the Yarra, is the source of Melbourne's largest reservoir, followed by the Upper Yarra reservoir and the Cardina, which contains a creek of that name running from the nearby Dandenong Range.

Like Cape Town, the water infrastructure came on a staggered basis as the population increased and periodic drought struck. The Upper Yarra was approved in the early 1940s, during an eight-year dry spell that lasted between 1937 and 1945, finally finishing construction in 1957 following a long delay as a result of World War II. This "World War II drought" saw a number of drought breaks with subsequent periods of extreme dryness. Beginning in New South Wales and Victoria, the drought spread further west, finding some reprieve in 1939 before collapsing again and emptying the Nepean Dam. After finding some rain again in 1941, the drought fell again, halting the flow of Australia's largest river, the Murray, to wetlands and flood plains by 1945 (LeBlanc et al. 2012). Amid worsening dust storms, the drought finally broke in 1945 (Sauter 2015).

Beginning construction in the 1960s, the Cardina was similarly prompted by a dry spell over the area between 1965 and 1968. During this period, Tasmania saw intense fires across the Bass Straight from Melbourne, while New South Wales vacillated between extreme dryness and rainfall from 1964 to 1966. The following year 1967 saw the driest six-month stretch of the twentieth century for Victoria and a return of dust storms in South Australia before the drought ended in 1968 (Barua et al. 2009).

The Thompson Reservoir—the most recent of the city's ten major storage facilities—was conceived to resolve lingering concern over water scarcity in the 1970s and was finished in 1983, at the end of a severe year-long drought. The 1982–1983 drought took place during a strong El Niño system and saw a significant lack of moisture, leading to hard frosts in the winter. Melbourne issued water restrictions, while inland in Victoria, authorities issued burn bans earlier than ever. Renewed Tasmanian brush fires and wildfires along with dust storms in Victoria marked the nadir of the drought in 1983 (McCulloch et al. 1994; McTainsh et al. 2005).

## 9.5 Climate Change and Drought

The previous droughts, which recurred at intervals of 15–30 years and lasted between 1 and 8 years, roughly matched climatological variability patterns, but the Millenium Drought, occurring between 1997 and 2009, was the first that involved the impacts of climate change. Stretching even longer than the World War II drought, the Millenium Drought spanned a long period of years without above-average rainfall, despite a few years of extreme dryness. In general, this period was defined by three strong El Niño events, causing high-pressure ridges to block out storm systems. In particular, autumn rainfall declined, bringing about drier conditions around rivers and lower storage levels (Van Dijk et al. 2013).

This long-term trend parched areas of the Murray-Darling Basin and crucial cropland. The Murray River fell to the lowest levels on record, capable of continuing low flows only through major irrigation sacrifices. Water use by irrigation declined to a third of its previous amount, while the gross value of agricultural production in irrigated areas dropped to just 80% in price-adjusted terms (Kirby et al. 2012). Meanwhile, the drought was punctuated by record breaking temperatures. This phenomenon posed unprecedented challenges to the country likely to be repeated in the future.

The Millennium Drought also significantly impacted Victoria. The area was hard hit from the beginning, and by the end of the drought, Melbourne's water storage reached just 25.6% of capacity (Low et al. 2015). While some areas experienced drier cold seasons and some rainy warm seasons, Tasmania faced hot and dry temperatures throughout the period. On top of these drier conditions, Melbourne saw a simultaneous phenomenon of brush fires and extreme temperatures, climbing up to 46.4 °C, the highest daily temperature on record, just weeks after the longest stretch of consecutive days above 43 °C (Guo et al. 2021).

Amid the highest temperature day, the brush fires caused 173 deaths and devastated over 2,000 houses on Saturday, February 7, 2009, earning the name Black Saturday. As many as 113 of the dead perished in their homes, adding to concern that Australia's preparatory framework for disaster-motivated evacuations lacked sufficient guidelines. "Climate research indicated a strong relationship between the rainfall decline in south-eastern Australia and the rise in intensity of the Sub Tropical Ridge (STR), the area of high-pressure systems over the region," according to a study authored by Bruce G. Rhodes and others. "This strengthening of the STR is estimated to account for around 80% of the rainfall decline in south-eastern Australia (Timbal 2007). The STR has intensified with the increasing global surface temperatures and implies that the rainfall decline may have some link to global warming" (Rhodes et al. 2012).

Following this extensive drought period, another impactful two-year drought occurred between 2017 and 2019, compounding drier existing conditions in Western Victoria. Again, dryer cool seasons did not see a reprieve during the warm season, leading to dryness in the Murray-Darling Basin that toppled the record set during the 1960s. Eastern Victoria and the eastern part of Tasmania were similarly hard hit.

Part of the issue in 2019 was an incredibly strong Indian Ocean dipole event, along with a negative Southern Annular Mode, which resulted in the eastward flow of dry air over Australia. As a result of the combination and intensification of climate variability, the droughts caused enormous and devastating wildfires during its final months (Nguyen et al. 2021). These fires claimed the lives of four people, along with some 900 structures, including over 350 homes in Victoria. In total, 1.4 million hectares were subsumed in the blaze, making Melbourne's the worst air quality in the world at that time.

Risks associated with drought include the increase of brush fires during dry, warm years. Fires can pollute waterways with ash and debris, while also destroying groundcover, causing subsequent erosion and siltation. Since Melbourne's water supply derives from surface water reservoirs, there is an issue with ash spreading into and contaminating the water supply. Also, streamflow reduction will lead to hazards for biodiversity in riparian ecosystems as thresholds for temperature maximums are crossed, and pollution rates increase relative to the amount of water left in-stream. Shortages can also lead to trade-offs for agricultural withdrawals, causing political fissures to expand.

## 9.6 Flooding in Melbourne

Between drought phases, the Melbourne area has experienced flood events. From the encroachment of moderate snow cover over the City of Melbourne to the advancement of deadly urban floods, the area's flood-drought cycle poses enhanced problems for managers and civilians, alike. These floods can come from pluvial nuisance floods, fluvial riverine flooding, or storm surges, compounding the complexity of flood hazards. With the rise of sea levels around the world, Melbourne stands to see such flooding worsen as time magnifies the effects of climate change.

Strangely, after the brutal World War II drought, Melbourne found the streets of downtown's business district covered with a layer of snow, present even in some areas of the suburbs, followed after a few years by torrential flooding and landslides. The flooding damaged shops in the central city, while suburban neighborhoods on the beach faced evacuation (Rogers et al. 2020).

Following the dire drought of the late 1960s, a large wave of floods struck Melbourne again in 1972, with 75 mm of water falling over 17 min of rain. The scene of cars floating down Elizabeth Street in central Melbourne caused a local spectacle. Five years after that, the suburb of Laverton faced a thunderstorm that lasted a full 12 h, dumping 395 mm and breaking regional rainfall records.

Again, the 1982–1983 drought was washed away the following year by a massive storm in the suburbs of eastern Melbourne that damaged more than 100 homes (The Sydney Morning Herald 1984). The next major flooding incidents came in the 2010s, with large events occurring in 2011, 2012, and 2018, when the city received 20mm of water in just 10 minutes (Travers et al. 2018). While the famous 2011 flood was influenced by a tropical cyclone that dumped more than 40 mm of rainfall on over a third of Victoria (with 15 stations reporting more than double that), leading to evacuations and almost $126.5 million in damage (Callaghan 2021).

According to climatologist Jeff Callaghan's study on the Southern Oscillation Index (SOI), climate change may mean that droughts and flooding have not corresponded as clearly to La Niña (positive SOI) and El Niño (negative SOI) events, respectively, as they have in the past. "The 1976 Climate Shift and its influence on significant Victorian rainfall events is studied and negative SOI monthly values were shown to dominate following the Shift," he wrote, indicating that El Niño patterns have dominated, which would typically mean more flooding for much of Australia. "However, one of the most active periods in 144 years of Victorian heavy rain occurred after the shift with a sustained period of positive SOI events from 2007 to 2014. Therefore, it is critical for forecasting future Victorian heavy rainfall to understand if sequences of these positive SOI events continue like those preceding the Shift" (Callaghan 2021). Most recently, the major flood of June 2021 struck Melbourne by way of intense rainfall accompanied by high river levels and fast winds. Evacuations occurred along nearby Traralgon Creek, as creek levels jumped two meters above moderate flood levels (ABC Gippsland and Fields 2021).

Flood risks go beyond traditional concerns about housing and crop damage or loss of life, as well. During droughts, increased sewage concentrations cause corrosion to pipes and worsened smells, as well as pipe collapse. Floods can cause overflows and damage to pump stations or underground drains, both of which can ultimately lead to contamination. In addition, sea level rise can bring about an increase in salinity in water that is recycled through the system, increasing the chances of corrosion as well as infiltration into water treatment facilities.

## 9.7 Water Sensitivity

As a result of the floods and droughts experienced in Melbourne, the city has taken up a "water-sensitive" approach. In 2004, the State of Victoria released a white paper called *Our Water Our Future: Securing Our Water Future Together*, calling for water conservation, increased efficiency, and restoration of watersheds (2012). The goal of these imperatives was not the implementation of a top-down, technocratic plan to resolve water issues through command-and-control but to build communities supportive of water sustainability from the bottom-up, supplemented by accountable water agencies.

The year 2004 saw the eighth year of the Millenium Drought, and the urgency for long-term reassessment has scarcely been higher. "In the future, Victoria is projected to get drier. Climate change is predicted to create more hot days, more dry days and more storms. This compounds our need to be smarter with available water supplies," the report said. "If Melburnians continue to use water in the future at the same rate and in the same way as they have in the 1990s, the city may approach its supply limits within 15 years." Confronting these alarming realities would take decisive action shared by the whole population, targeting the following:

- Optimization of sanitation and urban water supply
- Efficient irrigation systems
- Ecologically healthy rivers and streams
- Low-impact residential water use

The targets would necessitate striking a balance between the water needs of different communities, however. In order to achieve a sustainable balance, the report argued for the integration of stormwater and recycled water into the state's water system. Along with reforming the sellable water rights system, the report protected non-rechargeable aquifers, enhancing watershed protections, while also expanding the water monitoring capabilities of the state. "Improving the health of Victoria's rivers and wetlands requires an integrated approach dealing with all the major causes of river stress," the report stated, further noting that the water supply for Melbourne relied on the ecological destruction of the Thomson and Macalister rivers, necessitating restoration and monitoring in order to both provide water to the city and ensure native fish habitat.

Ideating creative solutions for water management, such as the replacement of riverine irrigation water with recycled urban water, brought Melbourne and the State of Victoria to the forefront of the international struggle to secure water resources in the face of climate change. However, some would argue that the policy framework had not yet been fully formed. The report lacked "environmental protection aspects of the urban water cycle," according to one critic, and its calls for reform did not address urban development as a pressure on water supply (Wong 2006). Whether or not such criticisms could be viewed as uncharitable, the report offered a basis for further developments in water resources management, including the early implementation of a "water-sensitive cities" framework in Melbourne.

Here, rather than focusing on irrigation and restoration upstream, along with source-based conceptions of water system reforms, the water-sensitive urban design approach facilitated the creation of household water budget solutions. These efforts seek to "close the loop" in water waste and need, sending greywater (from washing up, for example) to a basement treatment plant to be then pumped into the toilets for second use. Similarly, roof runoff can feed roof gardens and rainwater cisterns subsequently integrated into the household system. On a larger scale, stormwater treatment is integrated into localized treatment and bio-retention designs, such as miniature wetlands, swales, and rain gardens, which keep water for infiltration rather than overland flow (Wong 2006).

These types of measures approach urban issues from a socio-technical position in accordance with a multilevel model by bringing institutional regimes together with large-scale social systems and the level of technical innovations. While the Millenium Drought can be seen as the origin point of integrated and water-sensitive thinking in Australia from a policy level, the large-scale shifts in social and cultural thinking relative to the environment came to the fore as far back as the 1970s and 1980s. By the 1990s, this way of thinking brought about resources, and the environment could be worked into new legislation through newly established institutions. Urban stormwater and watershed restoration had been increasingly implemented at the onset of the Millennium Drought, and subsequent developments such as the first biannual National WSUD Conference in Melbourne in 2000 could bring together scientists and policymakers to craft a conscientious response to climate-induced water issues, build capacity, and establish adaptive learning strategies (Brown and Clarke 2007).

While these water-sensitive urban design (WSUD) practices appeal to water managers in times of crisis, attempts to evaluate their implementation have been few and far between, and there is little agreement on methods to assess the success or failures of the program. For instance, a study of WSUD at the University of Melbourne, where one might expect the most robust implementation, found that "the existing stormwater tanks are grey-based WSUD facilities, and the green-based WSUD facilities such as bioretention areas and swales are very few. For these reasons, it is far from achieving the sustainable storm" (Xiong et al. 2020).

Another study on the implementation of WSUD notes that the practice hinges on the involvement of municipal areas, which are not always motivated to invest time and resources toward such ends. "While municipal councils are recognised as vital collaborators for the improvement of urban stormwater, the observed variability and often lack of municipal commitment to intergovernmental stormwater policy creates a distinct problem for achieving better ecological outcomes." Indeed, the most enthusiastic supporters of WSUD seemed to be the more wealthy seaside communities rather than the bulk of the population (Morison and Brown 2011). According to the study's authors, one explanation of this problem is that WSUD "is a collective action problem with limited incentives for public action: the catchment and waterways health problem is considered remote; policy responses and actions are perceived to be relatively costly; and the benefits from responding are 'sufficiently diffuse to preclude individualised action'" (Morison and Brown 2011).

In addition, a study of the spatial distribution of WSUD indicates some unfortunate problems in planning. For instance, WSUD features are often placed at stormwater drainage outlets rather than at the source, which would provide better flood control and pollution solutions (Kuller et al. 2018). This could be a result of wicked problems in urban development, miscommunication between planners and water managers, or issues with building and development codes. However, groups such as the Cooperative Research Center for Catchment Hydrology are working on nonpoint source pollution reduction through nonstructural as well as structural solutions that combine the best technological and design practices (Wong 2001).

At the same time, the placement of WSUD did not correlate to housing prices or distance from the city center. So, while the wealthy tended to support WSUD more, according to one study, WSUD does not benefit the wealthy disproportionately. While on one level, the lack of inequitable water management seems reassuring, on the other hand, it can prove problematic. "A potential lack of understanding and appreciation for WSUD, resulting from low environmental awareness and education levels, may cause a lack of acceptance and intentional and unintentional maltreatment of these assets, jeopardising their operation" (Kuller et al. 2018).

## 9.8 Long-Term Design

In 2012, the state created an Office of Living Victoria in order to establish a path toward long-term sustainability. Its chief scientist, a controversial academic named Peter Coombes, advanced strategies in line with WSUD, calling for water recycling, bioswales, and other ways of conserving water resources. Viewed as an alternative to desalination plants or piping irrigation water to the city, the Office of Living Victoria was supported with $50 million in state funds to set its ideas into action. Yet, the difficulty of WSUD to obtain a popular foothold in rural areas brought a lack of popularity and concerns that taxpayer money was being spent on pipe dreams.

Despite reservations, plans to set forward a strategy received new contributions from optimistic fellow travelers. The year after the creation of the Office of Living Victoria, scientists from Monash University in Victoria and the Dutch Research Institute for Transitions partnered with Erasmus University in Rotterdam, publishing a study in *Landscape and Urban Planning*. Declaring that "complexity, variability and uncertainty will characterize urban water futures and conventional water planning is inadequate to deliver solutions that will cope with this context," Ferguson et al. (2013) identified a sensible long-term strategy for transitioning Melbourne to a water-sensitive city. Using document analysis and focus groups to assess the imperatives for transition, the team developed a system analysis of urban water driven by the guiding principles of long-term resilience to climate change and producing storylines for a desirable future. Formulating an agenda that prioritizes tasks in accordance with popular needs, the scientists synthesized the underlying challenges with a future vision in a strategic pathway forward. Through that process, the study created a strategic program deemed, "a radical departure from the traditional centralized infrastructure system" (Ferguson et al. 2013).

The study outlined a desired vision shared by workshop participants of lavish greenways, clean water, and savings brought back to consumers. Yet, the problems described in the study remained significant. The lack of commitment by political parties, the lack of a compelling vision, the confusing array of relationships being brought together, the legacy of past failures, the limitation of economic revenue, and the lack of capacity all stood out. The strategic transition pathways that the study's authors resolved to follow, in light of those problems, seemed ambitious to say the least: "Embedding" the shared vision would take time, fostering community would prove difficult, developing collaborative frameworks would require a social base, while supporting innovation and harmonizing planning with water values would need investment (Ferguson et al. 2013).

Indeed, the following year, the Office of Living Victoria came under fire for spending millions of dollars on consultants with past ties to Coombes as well as chief executive, Mike Waller and head of office, Simon Want. To wit, Want and Coombes, themselves, received contracts amounting to some \$1.5 million. These irregularities stood out from traditional procurement processes and raised more than a few eyebrows. The Office's promise of providing \$7 billion in savings by 2050 started to come under scrutiny, as the leadership refused to peer-review its models or studies (Baker and McKenzie 2014).

The Labor Party candidate Daniel Andrews made the abolition of the Office of Living Victoria a campaign pledge, which he fulfilled in December 2014, marking the unfortunate end to an ambitious but opaque organization plagued by scandal. According to an Ombudsman investigation, the Office of Living Victoria breached procurement standards and governance protocols, pursuing a singular track without gaining public support or adhering to standards of accountability (Victorian Ombudsman 2014). Thus, the project was brought to an unseemly close, casting a pall over the illustrious ideas underpinning WSUD.

## 9.9 Conclusions

With an eye to the *Landscape and Urban Planning* study, it appears that water-sensitive infrastructure remains a key part of Melbourne's approach to dealing with water resources.

Yet, it is only one tool in a larger toolbox developed by stakeholders, including such efforts as the Climate Change Management framework. Here, the focus of water resource management involves ranking climate-related risks, mitigating emissions, and supporting further climate research. In addition, larger projects include not just a focus on integrated water resources management but developing the range of reservoirs to ensure access even at low levels, revising low-flow requirements to ensure public health, constructing an intra-basin pipeline, building a desalination plant, planning out a drought response that includes water restrictions, and implementing general water conservation (Rhodes et al. 2012).

These larger-scale projects remain contentious for their apparent top-down approach and the tendency to sacrifice ecosystem needs in the event of drought. In the words of one study published in *Environmental Research Letters*, "Future augmentation of the desalination plant may mitigate these concerns but may come at a substantial cost to the community and the environment." However, the same study finds that as climate change increases temperatures from 1.5 to 2°C above preindustrial levels, the risk of water shortage increases more than double for Melbourne. The study concludes, "Regardless of the success or failure of international climate negotiations, societies located in climate zones prone to concurrently warmer and drier conditions would be wise to remain vigilant in fortifying their water resource defences against reductions in precipitation and the exacerbating influences of ongoing warming of the land surface" (Henley et al. 2019).

Melbourne thus provides a case study for two separate phenomena: the impacts of climate change on flooding and drought and efforts to mitigate those impacts by instituting a water sensitive-urban design approach. While Melbourne's experiment in the latter suffered significant setbacks loaded with political contestation, it suggests that such a decentralized approach requires a serious commitment to the principles that it forwards, which may become more difficult as institutionalization and its incumbent bureaucratization start to occur. While water managers contend with the pressing issues of climate change, however, such decentralized efforts may offer the clearest way to protect biodiversity and ecosystem health while conserving water and preventing unwanted emissions or stakeholder sacrifices. At the same time, the problems of stakeholder commitment and acceptance remain as significant of a hurdle to WSUD as administrative scandals. Yet, through effective leadership and management, the marriage of complexity and water resources management remains feasible and likely more desirable than the externalities of large-scale, environmentally compromising water projects.

## References

ABC Gippsland, Field E (2021) Emergency authorities apologise for late flood warning, issued 1.5hrs after Traralgon Creek peaked. ABC News, June 14. https://www.abc.net.au/news/2021-06-15/traralgon-creek-flood-apology-gippsland/100213276

Baker R, McKenzie N (2014) Coalition water consultants Peter Coombes and Simon Want given top jobs in the Office of Living Victoria. The Age, February 18. https://www.theage.com.au/national/victoria/coalition-water-consultants-peter-coombes-and-simon-want-given-top-jobs-in-the-office-of-living-victoria-20140217-32wfc.html

Barua S, Perera BJC, Ng AWM (2009, July). A comparative drought assessment of Yarra River Catchment in Victoria, Australia. In: Proceedings of the 18th World IMACS/MODSIM Congress, Cairns, Australia, pp 13–17

BoM (2021) What is El Niño and how does it impact Australia?, Australian Government. Bureau of Meteorology, Melbourne

BoM (2023) Climate Data Online. Bureau of Meteorology, Melbourne. http://www.bom.gov.au/climate/data/index.shtml

Briggs CM (2019) Yulendj Boonwurrung: A journey of old knowledge and innovative forms for assisting urban Indigenous youth to engage in contemporary Indigenous knowledge. Doctoral dissertation, RMIT University

Brown R, Clarke J (2007, June) The transition towards Water Sensitive Urban Design: a socio-technical analysis of Melbourne, Australia. In: Novatech 2007-6ème Conférence sur les techniques et stratégies durables pour la gestion des eaux urbaines par temps de pluie/ Sixth International Conference on Sustainable Techniques and Strategies in Urban Water Management. GRAIE, Lyon, France

Callaghan J (2021) A climatology of heavy rain and major flood events in Victoria 1876-2019 and the effect of the 1976 climate shift. J Geograph Res 4(3):12–33

Ferguson BC, Frantzeskaki N, Brown RR (2013) A strategic program for transitioning to a Water Sensitive City. Landsc Urban Plann 117:32–45

Gibson JW, Pierce MC (2009). Remnants of early hydraulic power systems. 3rd Australasian engineering heritage conference

Guo Y, Zhang L, Zhang Y, Wang Z, Zheng HX (2021) Estimating impacts of wildfire and climate variability on streamflow in Victoria, Australia. Hydrol Process 35(12):e14439

Hameed SN, Jin D, Thilakan V (2018) A model for super El Niños. Nature Commun 9(1):2528

Henley BJ, Peel MC, Nathan R, King AD, Ukkola AM, Karoly DJ, Tan KS (2019) Amplification of risks to water supply at 1.5 C and 2 C in drying climates: a case study for Melbourne, Australia. Environ Res Lett 14(8):084028

Hong L-C, LinHo, Jin F-F (2014) A southern hemisphere booster of super El Niño. Geophys Res Lett 41(6):2142–2149

Howe C (2007) Implications of potential climate change for Melbourne's water resources

Jose J, Jackson L (2022) Australia suffers flash floods in southeast, Melbourne suburb evacuated, Reuters

Katz RW (2002) Sir Gilbert Walker and a connection between El Niño and statistics. Statist Sci 17(1):97–112

Kirby M., Connor JD, Bark RH, Qureshi ME, Keyworth SW (2012) The economic impact of water reductions during the Millennium Drought in the Murray-Darling Basin (No. 423-2016-26978)

Kuller M, Bach PM, Ramirez-Lovering D, Deletic A (2018) What drives the location choice for water sensitive infrastructure in Melbourne, Australia? Landsc Urban Plann 175:92–101

Leblanc M, Tweed S, Van Dijk A, Timbal B (2012) A review of historic and future hydrological changes in the Murray-Darling Basin. Global Planet Change 80:226–246

Low KG, Grant SB, Hamilton AJ, Gan K, Saphores JD, Arora M, Feldman DL (2015) Fighting drought with innovation: Melbourne's response to the Millennium Drought in Southeast Australia. Wiley Interdisciplinary Rev Water 2(4):315–328

Mariani M, Fletcher MS, Holz A, Nyman P (2016) ENSO controls interannual fire activity in southeast Australia. Geophys Res Lett 43(20):10–891

McCulloch MT, Gagan MK, Mortimer GE, Chivas AR, Isdale PJ (1994) A high-resolution Sr/Ca and δ18O coral record from the Great Barrier Reef, Australia, and the 1982–1983 El Niño. Geochimica et Cosmochimica Acta 58(12):2747–2754

McTainsh G, Chan YC, McGowan H, Leys J, Tews K (2005) The 23rd October 2002 dust storm in eastern Australia: characteristics and meteorological conditions. Atmospheric Environ 39(7):1227–1236

Morison PJ, Brown RR (2011) Understanding the nature of publics and local policy commitment to Water Sensitive Urban Design. Landsc Urban Plann 99(2):83–92

Neelin JD, Latif M (1998) El Nino dynamics. Phys Today 51(12):32

Nguyen H, Wheeler MC, Hendon HH, Lim EP, Otkin JA (2021) The 2019 flash droughts in subtropical eastern Australia and their association with large-scale climate drivers. Weather Climate Extremes 32:100321

Ore A (2023) Flooding of Melbourne retirement village left second world war refugee with PTSD, inquiry hears. The Guardian

Pierce M (2009) The Melbourne Hydraulic Power Company and public hydraulic power systems in Australia. Aust J Mech Eng 7(2):173–187

Rhodes BG, Fagan JE, Tan KS, Water M (2012, May) Responding to a rapid climate shift–experiences from Melbourne, Australia. In: IWA World Congress on water, climate and energy, Dublin, Ireland, 13–18 May 2012

Rogers BC, Bertram N, Gersonius B, Gunn A, Löwe R, Murphy C et al (2020) An interdisciplinary and catchment approach to enhancing urban flood resilience: a Melbourne case. Philos Trans Royal Soc A 378(2168):20190201

Sanders P, Lozanovska M, Van Galen L (2021) Lines of settlement: lost landscapes within maps for future morphologies. Heritage 4(3):1400–1414

Sauter S (2015) Australia's Dust Bowl: Transnational Influences in Soil Conservation and the Spread of Ecological Thought. Aust J Politics Hist 61(3):352–365

Slattery D (2015) Australian Alps: Kosciuszko, Alpine and Namadgi National Parks. Csiro Publishing, Clayton South

SWIFFT (2021) Threatened fauna Yarra Ranges Shire, State Wide Integrated Flora and Fauna Teams. https://www.swifft.net.au/cb_pages/threatened_fauna_yarra_ranges_shire.php

The Sydney Morning Herald (1984, September 19) Melbourne floods hit 100 homes

Timbal B (2007) Compare documented climate changes with those attributable to specific causes. South-Eastern Australia Clim Initiat 4

Travers B, Rose T, Hosking W (2018) Melbourne and surrounding suburbs cop a soaking in peak-hour downpour. Herald Sun, December 15. https://www.heraldsun.com.au/news/victoria/melbourne-and-surrounding-suburbs-cops-a-soaking-in-peakhour-downpour/news-story/dd3f259ff594dfc3bec9320a94536f46

Van Dijk AI, Beck HE, Crosbie RS, De Jeu RA, Liu YY, Podger GM et al (2013) The Millennium Drought in southeast Australia (2001–2009): natural and human causes and implications for water resources, ecosystems, economy, and society. Water Resour Res 49(2):1040–1057

Victorian Government White Paper(2004) Our water our future: securing our water future together

Victorian Ombudsman (2014) Investigation into allegations of improper conduct in the Office of Living Victoria

Walker GT, Bliss EW (1932) World weather V. Memoirs of the Royal Meteorological Society 4:53–84

Wong TH. (2001, June) A changing paradigm in Australian urban stormwater management. In: 2nd South Pacific stormwater conference, pp 1–18

Wong TH (2006) Water sensitive urban design-the journey thus far. Australasian J Water Resour 10(3):213–222

Xiong H, Sun Y, Ren X (2020) Comprehensive assessment of water sensitive urban design practices based on multi-criteria decision analysis via a case study of the University of Melbourne, Australia. Water 12(10):2885

# Chapter 10
# São Paulo, Brazil

**Abstract** São Paulo is one of Brazil's most economically productive and technologically advanced metropolises. Lying in the midst of a coastal mountain range but close to the coast, the megacity might not be prone to the worst aspects of coastal flooding, but it inherits significant challenges from a century of modern urbanization and efforts to contain its powerful river systems. This chapter examines the complex and multifaceted character of southeastern Brazil's climate system, including variability, showing the importance of flooding and droughts in population flux and urban water crisis. Discussing recent efforts, such as piscinões, to promote adaptive water management, it indicates that some major financial investments and adjustments are needed to bring down the perennial costs of flooding and drought, as well as their implications for São Paulo's poorest residents.

**Keywords** Piscinões · Channelization · Disappeared streams · River restoration · Hydro-basins

## 10.1 Introduction

It was the weekend of Carnival in São Paulo in 2023—a time of festivities associated with bright costumes, exuberant dances, and rituals of joy and abundance. However, there was a low-pressure system brewing off the coast. As partygoers filled the streets and tourists filed into their hotel rooms in the southeastern Brazilian megacity, a wet weather system rolled into the coastal area, gaining force with the orographic lift over the Serra do Mar range, accumulating into a violent storm that dumped some 30 inches in just 24 h in some areas. The flood event caused major landslides across coastal areas of the state, causing 64 fatalities. The Port of Santos was shut down, and the highway blew out amid 34 mph winds, and Carnival festivities were canceled across the coastal area. In the aftermath, horror was added to the devastation as looters took advantage of the chaos and humanitarian shipments were plundered.



H. Chang, A. R. Ross, *Climate Change, Urbanization, and Water Resources*, https://doi.org/10.1007/978-3-031-49631-8_10

The state of São Paulo is increasingly affected by climate change, be it in the form of debilitating floods or parching droughts, and its capital city is no different. Lying on the foothills of the leeward side of a majestic subtropical mountain range some 2500 feet above sea level, the city reflects the greatness of Brazilian modernism, as well as its drawbacks. The largest city in Brazil—in fact, in the western hemisphere—São Paulo holds the most skyscrapers over 115 feet tall in the world, with the towering Platina 220 built in 2022 reaching 564 feet, 50 floors high.

## 10.2 Geography

Contributing some 20% of the total GDP of Brazil, the metropolitan area of São Paulo is the fulcrum of Brazil's powerful financial industry and economy. The median household income in São Paulo is above the national average, and the city is home to the most active tech ecosystem in Latin America, home to the most start-ups in the country's fast-growing tech industry. But close to 20% of the city lives in poverty, with 1.2 million people living in *favelas* or illegally occupied buildings, called *corticos*. However, *favelas* are not the only places where poor people live. Nearly equal proportion of the bottom tenth people in São Paulo live in both favelas and non-favelas tracts (Carvalho and de Carvalho Cabra 2021).

Inhered within its rapid growth since 1950, when the city held merely one-tenth of its current population, deep dilemmas related to water issues were made all the more serious by climate hazards. Now part of the Atlantic Forest South-East Reserves, what remains of the autochthonous rainforests of what is now São Paulo tells a unique story of the particular evolution of the biotic communities that inhabit the Tropic of Capricorn. Thrown back against the dramatic backdrop of the Serra do Mar mountain range, which stakes a discontinuous path along the Atlantic coast from a 1000 km northeast all the way—some argue—to Brazil's southernmost Rio Grande do Sul, the southern Atlantic Forest comprises such immense biodiversity as to contain more than 450 tree species per hectare. In total, 120 separate species of mammal can be found among the 350 species of avifauna here, where the jaguar and the ocelot compete for armadillos with the native Caiçaras people—the descendants of Amerindians and European colonizers.

On lands inhabited by the indigenous Tupi-Guarani people who still dwell in discontiguous territories throughout the Amazon and in Bolivia, Uruguay, Paraguay, and northern Argentina, the village of São Paulo was established by Jesuit missionaries and Portuguese explorers. According to SOSMA/INPE (2018), 86% of the original forest has been converted to agriculture or residential development in the state of São Paulo (Southeastern Region of Brazil) since its settlement. Not only coffee plantations but also sugarcane and orange plantations came to dominate the verdant and lush region, where rains typically compensate for the lack of adequate infrastructure systems.

## 10.3 Climate

Its position high above sea level brings São Paulo an advantageous temperate climate, given its latitude at the tropical boundary. The colder month of July tends to an average of 58 °F, while the peak summer month of February usually levels out at 69 °F. Although seasonal variability is typically minimal in the tropics, in terms of precipitation, the wet months of the summer (9.5 in) differ starkly from the drier months of the winter (1.6 in). Indeed, seasonal variability in precipitation is marked by surprising variability in temperature as a result of the coinciding warm air masses overland, cool air from the ocean, and the Antarctic.

Extreme rainfall events in the state of São Paulo are generally characterized by the presence of higher sea surface temperature anomalies in the equatorial Pacific Ocean off the South American coast, whereas extreme events along the coastal area of the state correspond to warm anomalies in the near Atlantic. Rainfall events can also be enhanced by the presence of the South Atlantic convergence zone, an atmospheric trough that bends down from the equator and sinks into southeastern Brazil with the potential to unleash torrential downpours (Liebmann et al. 2001).

When the South Atlantic convergence zone is weaker, another system takes over, called the South American dipole. This system replaces the wetter trough with high-pressure systems off the southeastern coast of Brazil. This descending wind anomaly then converges further to the southwest, creating low-pressure conditions in southwestern Latin America that help circulate rains to Argentina and Uruguay (Seth et al. 2015).

In the words of Nobre and Marengo (2017), "In the Metropolitan Region of São Paulo city (MRSP), total rainfall has been increasing since 1961, but precipitation patterns are getting more irregular, so that very intense rainfall events are concentrated in short periods of time (i.e., a few days) that are separated by long periods of very hot and dry climate." As in other areas, storms may be fewer and further between, but they may also be more intense. Cases of precipitation over 30 mm/day increased by 40% from 1933 to 1940 to 2000 to 2009. Meanwhile, precipitation in the form of drizzle lower than 1 mm per day decreased over the last 20 years (de Lima et al. 2018).

## 10.4 River System

Extreme precipitation events can cause the Tietê River to overflow its banks, leading to flooding that can prove incredibly destructive. As coffee production expanded through the interior of São Paulo through the nineteenth century, the city's wealth grew, and with it, the process of industrial development grew. The city spread to the fringes of the Tietê floodplain, and those of its tributaries, Tramanduatei and Pinheiros, which flow through the city, itself. Initially on the city's border, the river floodplain became a makeshift residential zone for the poorest residents, along with leisure sports such as soccer (Fig. 10.1).

**Fig. 10.1** Landscape change along the Tietê River in São Paulo, (**a**) 1905 and (**b**) 2011. (Photo credit: Wikipedia, 2023)

A 12-million-year-old stream carving a northwestern path out of the crystalline rocks of the Serra do Mar, the Tietê winds and twists from Oscasco to Mogi das Cruzes, scouring out labyrinthine meanders as it discharges more than 88,000 cubic feet per second into the Paraná River and, from thence, south into the Rio de la Plata (James 1933). The power and threat that the river and its tributaries imposed on the growing city did not go unnoticed, of course, and city officials first opened rectification projects on the Tramanduatei in 1842. Yet, further action, including plans to bend the Tietê and Pinheiros into a circular perimeter around the city, went unfulfilled into the twentieth century.

Locals continued using the riverine areas for fishing and sports activities during the early twentieth century, as the Tietê River Improvement Commission developed plans to straighten and curve the river. In 1932, the elites of São Paulo, who believed that their economic might outstripped their political power, attempted to overthrow the government, resulting in a reorganization of administrative positions that brought engineering to the forefront of the political process (Peixoto-Mehrtens 2010; Millington 2021). As a result, state planning for public works accelerated, along with the technocratic means of implementing them in tandem with public–private partnerships.

A further plan produced in 1950 stemmed from the offices of none other than Robert Moses. However, river straightening finally went through in accordance with the plans of João Florence de Ulhôa Cintra, deepening the river by 4 feet and channelizing it through the metropolis. While the plan reclaimed floodplain areas for more sanitary urban development, power was the key use. São Paulo Tramway, Light and Power Company, a Canadian firm typically called Light, received the rights to generate two river dams. Unfortunately, the municipality hoped to keep the dams low in order to ensure an adequate flow through the river system, disposing of any unsanitary waste, the power company Light kept them full to maximize electricity generation. Light manipulated their position, and the promise of the land within the Pinheiros floodplain to grab as much land as possible by releasing their floodgates and subsequently causing a massive flood, which they could then claim as demarcating the extent of their property (Millington 2021).

Despite the murky ethical transactions, the channelized rivers that once formed barriers between the central city and its eastern (in the case of the Tietê) and the northern (in the case of the Pinheiros) margins now served as easily surmountable through the national road plan. The city grew, in turn, from one million residents in 1945 to 4.7 in 1960 and 8.2 by the 1970s, and the simultaneous growth of federally funded infrastructure led to the sustained development of impervious surfaces in the form of highways and stream rectification (Fig. 10.2).

Yet, the environmental impacts of channelization are serious. Siltation grew as massive amounts of waste poured into the rivers. Cleaning the system of debris proved expensive and environmentally deleterious to disposal sites. The growth of impervious surfaces further contaminates the river system with chemicals and contaminants washed from the streets of the city directly into the streams. With some five million vehicles, this means large amounts of oil and gasoline in the water. Furthermore, the increase in stream velocity caused by channelization contributes to more sediment deposits in-stream, which can lead to increased flooding. Thus, the channelization process is ongoing, requiring constant dredging to ensure channel depth and stave off the threat of floods (Pessoa 2019). The main tributaries of the Tietê River have steep slopes, i.e., these tributaries discharge their floods quickly into the Tietê, which is slower and cannot support large floods

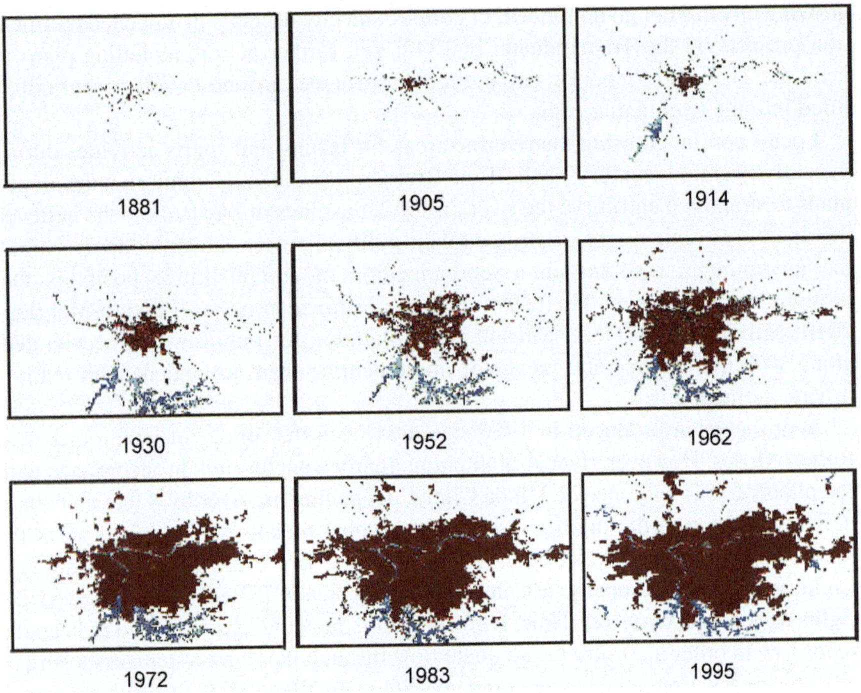


**Fig. 10.2** Land cover change in São Paulo, 1881–1995

(Dall'Acqua and Gertsenchtein 2013). The degradation of water quality, combined with increasing demand for clean water supply, stresses the water resource management of São Paulo.

## 10.5 Droughts

The most fearsome droughts in Brazilian history struck the northeastern area, with the worst event—the Grande Seca—taking place in 1877 during an El Niño year. In a traumatic incident, the drought killed some half-million people, displacing hundreds of thousands more. Periodic drought continues to plague the northeast, which remains among Brazil's poorer regions. The São Paulo region has experienced severe droughts in history, as well, and recent events suggest a trend toward heavier impacts and longer-lasting systems.

The most severe drought in the São Paulo region took place in the beginning of the 1940s during the administration of Getúlio Vargas. A repressive dictator who ruthlessly suppressed the labor movement while posing as the "father of the poor," Vargas faced two major droughts—one in 1932 and another hitting São Paulo in 1942. With the potential entrance of Brazil into the World War II, Vargas decided to take up a commanding control over the labor and food markets. The year 1941 had been dry, and the spring of 1942 showed signs of little improvement. As was by now the typical story, in the northeast, drought brought an influx of desperate migrants, large government work sites became sites of tremendous clashes, and rioters scoured the cities for food and loot, while thousands of flagellates flocked to the capitols. The government responded by herding refugees into concentration camps, keeping the flow of labor and production going under the supervision of technocratic authorities, simultaneously hoping generously to cultivate a consumer base and manipulate streams of government aid (de Castro Neves 2001).

The 1960s saw the next major period of extreme drought. In 1963, a drought initiated in eastern Brazil in April 1963 and spread north to Amazonia the following Spring, marking the worst 12-month drought. In June 1968, another drought began in the north and spread southward all the way to Paraguay (Sheffield et al. 2009), followed by another drought just 2 years later, largely afflicting the northeast.

Amid the difficult climatic conditions and sharp population growth, in part due to refugees from the northeast, São Paulo's state government began construction on the Sistema Cantareira in the foothills between 750 and 850 m in altitude. A system of reservoirs located at the headwaters of the Piracicaba tributary to the Tietê, the Cantareira began with a series of dams amounting to some 11,000 liters per second, with more reservoirs constructed again in the mid-1970s and tripling the initial amount.

In São Paulo, although a linear trend suggests a decrease in the severity of droughts over the past decades, as opposed to the rough 1940s and 1960s, three successive droughts struck in 1990, 2000–2001, and again beginning in 2012, showing that major crises can still emerge (Gozzo et al. 2019). The notable recent drought, which lasted 4 years from 2012 to 2015, exemplifies the current problems of large-scale mismanagement of water in the city. In 2013, rainfall had fallen under the historical average, showing the worst drought since 1930 in Brazil's southeast. The following year, water managers resorted to emergency draws from the 30 miles of pipes and six reservoirs composing the Cantareira system (Fontão and Ferreira 2022).

The first to feel the grip of the water crisis were irrigators and hydroelectric power operators. Safe drinking water to poorer districts in the hills around the city—its *periferia*—failed as well. The Cantareira system dropped to 12% of its normal capacity, exposing the dried mud on the riverbeds. Soon, hospitals and schools lost clean water, and the health of patients and young children began to decline. Jon Gerberg of *Time* describes the disastrous conditions.

> Many São Paulo residents have had daily 12-hour water cutoffs over the last year. But Dallari points out that while wealthier residents have been able to build water tanks and purchase water from private sources, the poorest residents can't do that… In the south of the city lies the Billings reservoir, which holds 20% more water than the Cantareira. Environmentalists point out that this could be a better source of water for the city, but it is polluted. Over one million people now live by its banks, but there is no proper sewage system so waste flows into the reservoir (Gerberg 2015).

Despite the fact that extreme drought in São Paulo has lessened somewhat since the drought of 1930 and the following decade, the calamity of 2015 showed water managers that water scarcity was, in fact, a major problem for the entire municipal area. de Lima (et al.) wrote "The [Metropolitan Area São Paulo] presents today one of the most critical situations in the country with regard to ensuring sufficient water supply in quantity and quality for its population." "Declining water resources coupled with increased demand for clean water has already become a political issue in many localities. This situation is becoming more and more aggravated by the constant urban expansion that generates the occupation of peripheral spaces and environmental protection areas, where are located drainage headwaters and important water bodies" (2018).

Enhancing the problem of scarcity, de Lima notes, is the irregularity of reservoir maintenance. He wrote that "Even in similar climatic conditions, some reservoirs had a high volume of stored water, while others oscillated rapidly, often exhibiting values well below their capacity" (2018). This problem of water management increases the stress on water supply, while also contributing to water quality issues with higher concentrations of pollutants in lower-volume storage and higher temperatures that can deplete dissolved oxygen.

## 10.6 Floods

The impacts of droughts can be felt, particularly, in relation to urban development, as density enhances the heat, also leading to more evaporation. At the same time, in less dry periods, that "urban heat island" effect can reinforce trends of rainfall events, contributing to flooding, which, even more than droughts, bring major challenges to São Paulo. Every summer, flooding strikes areas of the city, impacting regional residential life and global economic activity.

Primarily, the development of residential areas in the Tietê floodplain leads to persistent drainage problems and surplus runoff. As in other cities, the imposition of impervious surfaces through urban development gives water no place to go, and the basin morphology contributes to the population of low-lying areas, which inevitably face the worst of the flooding. Channelization only controlled the direction of the streamflow of the Tietê and its tributaries; it did not prevent overflow and, in fact, contributed to it by narrowing the area that the stream can fill. According to Kaźmierczak and Cavan (2011), poorer, underserved parts of urban populations are the most vulnerable.

In São Paulo, health risks remain very serious for such populations. The Tietê River is so polluted with thick toxic foams formed in parts of the river, which results from illegal untreated wastewater discharge. Water pollution has increased by 40% in 2022 (CNN 2023). Also, the average commute of someone in São Paulo nears two and a half hours, much of which takes place in cars. Periodic floods only expand that duration, leading to increased congestion, idling cars, traffic accidents, and emissions, along with the deposition of diesel fuel and other contaminants from the roads into surrounding areas and streams (Tomás et al. 2022).

Incredibly, a full 64% of surface water flooding occurred on São Paulo's arterial roads (Tomás et al. 2022). While the development of dense residential places in flood-prone areas has devastating consequences for socially vulnerable people, São Paulo's economy is also directly impacted by flooding, increasing the hazards for people even outside those areas. According to Haddad and Teixeira (2015), flood events can lead to more than a hundred billion dollars of loss to the city's GDP, with impacts reverberating throughout the country. The bigger the area impacted by floods, the more costly the impact is in terms of delays in services, production, and transportation.

Like Jakarta, São Paulo's failure to contend with flooding can lead to broader disillusionment with government agencies (Quintslr et al. 2021). Disillusionment can draw people away from reporting flood problems, leading to more ineffective flood control responses, and can reinforce the tendency of informal settlements to proliferate outside of well-regulated, healthy, and safe areas. As a result of the perpetuation of informal settlements, unsanitary and unsafe conditions can proliferate until the next flood destroys the area, leading to more pollution in the river and the loss of important dwellings.

As with all case studies examined, developing urban impervious surfaces continues to worsen pluvial flooding and impact more marginalized and vulnerable groups disproportionately. Young and Papini (2020) used a hydrologic model to identify the extent of flood exposure in the slums of the city, finding more than 5000 at-risk residents. Recent studies have also shown that "disappeared streams"—streams that have been paved over or otherwise filled in—tend to flood more easily, indicating an important issue for São Paulo's planners (Post et al. 2022).

## 10.7 Adaptive Strategies

As shown above, Brazil's climate faces a number of important challenges with regard to water issues. Dry periods bring higher demand for water even as the water supply is threatened, while the problem of untreated sewage remains high (ANA 2015). To confront the climate-enhanced problems of extreme rain events and dry periods, São Paulo is taking an adaptive approach that deploys a number of different strategies. Yet, these strategies have critics who say the focus on top-down solutions remains problematic.

One response of Brazilian planners to flood control issues is the structural feature of piscinões (Fig. 10.3). Generally made of concrete, these "pools" are dug into the earth up to 20 meters' depth and connected to the municipal stormwater sewer system. Water from intense rain fills the piscinões, removing it from the stormwater system during the rain event and pumping it back into the stream afterward. Some piscinões are buried under the surface, while most remain open-air.

First produced in 1994, the piscinões are part of a broader, state-led effort in which the province's Department of Water and Electrical Energy, together with the Secretary of Urban Infrastructure and Public Works, produces lucrative contracts

**Fig. 10.3** A major piscinão in São Paulo. (Source: São Paulo City 2017, Photo credit: Cesar Ogata/Secom)

for private firms generally in the engineering and construction sectors. Beginning in the 1990s, the piscinões represented one way of addressing the difficulty of removing surplus surface water from the urban area by implementing a new strategic paradigm to find a place for water to go within the city. São Paulo's *Master Plan for Macro-Drainage of the Alto-Tietê Basin* called for a stop to the deepening of the river, leaving the alternative to retain water within the municipality through landscape features. Using examples of Sustainable Urban Drainage Systems from Bordeaux, France, the piscinões creates a kind of alternative, urban floodplain, intending to "delay" rather than assimilate the surface water caused by storms (Millington 2021).

Since the first piscinão, a discrete, underground structure in a wealthy district, they have become larger and more imposing. Created in 2017 for \$30 million, the Guamiranga piscinão has a capacity of up to 850,000 $m^3$. Such an enormous and expensive structure, covering 70,000 $m^2$ of area, obviously rankles supporters of smaller-scale efforts that integrate design with values such as livability and green space. Critics insist that the influence of funding and the legacy of concrete development provide perverse incentives to develop gigantic structures such as freeways and piscinões, rather than address localized problems with less construction-intensive solutions. Often located in poorer, marginalized areas, piscinões require the displacement of residential developments, reconstruction of roads, and transformation of neighborhoods (Millington 2021).

On top of the implementation of piscinões, their maintenance also creates a problem. Residents complain about illegal dumping in piscinões and the failure to keep them clean. Waste deposited in piscinões creates both water quality problems downstream and local sanitation hazards in situ. Aside from water contamination and aesthetics, piscinões tend to attract mosquitos, which bear diseases that can cause major problems for local residents. The problems that piscinões create, then, can become costly externalities of one-and-done projects that, while pricey in the short term, neglect other aspects of the flood prevention budget, which is perennially underspent (Millington 2021).

At the same time, other adaptive strategies are taking root in São Paulo to more effective ends. One alternative strategy involves the reforestation of areas of the São Paulo municipality to improve water quality and reduce the quantity of runoff into the urban area. This effort is relatively new and still in the evaluation stage; a recent study (Ferreira et al. 2021) found that reforestation would greatly improve water quality by reducing the erosion of the soil and the kind of pollution associated with impermeable surfaces. Aside from its simultaneous benefits for the climate, reforestation also increases water quality by allowing infiltration into aquifers. These contributions to water quality and quantity only become more important with the encroaching challenges of climate change.

With respect to droughts and periods of water scarcity, some in São Paulo have suggested reclaiming the "hidden water resources" of the city by "mapping the network of rivers and natural springs that have been buried by buildings, roads, and other infrastructure." A local activist, Adriano Sampaio, promotes the idea of unearthing such "disappeared streams" through a Facebook page translatable as "There is Water in the City" (Gerberg 2015). It would stand to reason that, in restoring some streamflow to these developed areas, more overflow areas would be available for stormwater discharge during the rainy season, as well as alleviating water quantity issues during dry periods.

Indeed, restoration efforts are taking place in São Paulo. The Ciudade Azul project reveals a map of underground water, posting signs along streets to show the way for pedestrians hoping to travel through the winding hydrologic network that once made up the territory. But still more work is being done to control pollution, take out irrigation withdrawal points, replant native species, and destroy the concrete passages that force the flow of the stream. By demolishing streams' present structures, scientists hope to reintroduce former meanders and develop new ways to recreate the original function of stream behavior. While such interventions can be costly, if endeavored in the right places, they can make a substantial amount of difference to local flora and fauna, quality of life, ecological health, and water quality.

Alencar da Silva and Ferreira do Amaral Porto (2019) offer excellent advice on the future of stream rehabilitation in the city. One such stream is the Itaim tributary of the Jaguaré, which runs through Tizo Park in the city's westernmost area, a protected stream that still receives pollutants from nearby residential areas and would provide an ideal place for restoration from its present canal form. Further proposals for restoration include the buried stream underneath the Squares Evandro Valério and Padre Campos—these areas have scenic potential, but the "Spring stream" that runs underneath remains similarly influenced by residential pollution, whereas restoration would improve the water quality as well as the urban heat island effect.

In addition to these aspirational restoration projects, the City of São Paulo incorporated into their master plans during the early 2000s the intention to create "Linear

Parks," described as "a continuous system of green areas along valley floors, which have for objectives to preserve and restore the natural environment and natural water drainage and retention [functions], in addition to configuring public space for leisure and mobility" (Municipality of São Paulo 2021). Also in the Jaguaré basin, São Paulo implemented such intentions by developing green areas along waterways located amid informal settlements. In the words of Diep et al. (2022), "achieve stream restoration, provide services to informal areas, regenerate the wider neighbourhoods in which they are located, but also improve integration of these urban areas into the wider city fabric."

According to the Diep team's surveys and focus groups, the linear parks took a long time to develop and followed a staggered and unpredictable timeline. However, they appear to have decreased flooding problems while increasing the water quality in the short term. While wastewater disposal problems increased, the water quality declined and promised housing was never developed, so shacks returned to the same areas from which the state had evicted families. As the projects remained incomplete and poorly maintained, they fulfilled some but not all of their purpose, leaving or even worsening many of the issues they hoped to resolve (Diep et al. 2022).

## 10.8 Conclusions

São Paulo's efforts to reckon with flooding and droughts offer a synthesis between participatory and top-down water management paradigms. The production of piscinões, the unearthing of buried streams, and the creation of linear parks suggest variants on both nature-based and sustainable urban drainage systems. In different ways, they can resolve and create problems, and both illustrate the problems of top-heavy water management.

Though the piscinões help remove water from the surface, they create water quality hazards, lead to mosquito infestations, and uproot communities. Similarly, the nature-based restoration of streams promotes flood protection but, left unmanaged, also brings water quality problems and community neglect. Due to the importance of the crises brought by climate change, both efforts are useful and needed in their own ways but would be improved through greater input from residents.

For instance, piscinões could be developed to include plants with extensive root systems that would infiltrate rainwater into groundwater flow, rather than using concrete that enables pooling. Also, stream redevelopment and rehabilitation require extensive cleaning operations—some of which have been enacted in recent years to great effect. Furthermore, reforestation efforts have increased soil permeability and water quality in the Tietê basin. Regular maintenance and deepening of these projects, along with more equally distributed efforts throughout the city, could contribute to more manageable hydrologic conditions both in times of flood and drought.

De Lima (2018) notes that aside from the creation of "hydro-basins" to move water from a stressed area to other areas, the recycling of water can help recharge groundwater, while harvesting rainwater can improve problems of scarcity. São

Paulo already instills the need for rainwater harvesting and promotes water responsibility, particularly in larger buildings, but water recycling could be integrated with restoration and removal efforts to increase efficiency. Lastly, urban water infrastructure needs to be updated, particularly as it pertains to sewer pipes, in order to repair, upgrade, and remove pipes that need work or have become redundant.

## References


Alencar JC, do Amaral Porto MF (2019) Restoring, revitalizing and recovering Brazilian rivers: application of the concept to small basins in the City of São Paulo, Brazil. International Journal of Urban and Civil Engineering 13(3):183–189

ANA – Agência Nacional de Águas, 2015. Encarte Especial sobre a Crise Hídrica. Conjuntura dos Recursos Hídricos no Brasil: Informe 2014. Brasília (30 p)

Carvalho C, de Carvalho Cabra D (2021) Beyond the favelas: an analysis of Intraurban poverty patterns in Brazil. Prof Geogr 73(2):269–281. https://doi.org/10.1080/00330124.2020.1844571

CNN (2023) River is so polluted that it now has layer of toxic foam, July 8, 2023

Dall'Acqua C, Gertsenchtein AS (2013) Megaprojects in the São Paulo metropolitan region. Advice from those who've been there, done that, 279

de Lima GN, Lombardo MA, Magaña V (2018) Urban water supply and the changes in the precipitation patterns in the metropolitan area of São Paulo–Brazil. Appl Geogr 94:223–229

Diep L, Parikh P, dos Santos Duarte BP, Bourget AF, Dodman D, Martins JRS (2022) "It won't work here": lessons for just nature-based stream restoration in the context of urban informality. Environ Sci Pol 136:542–554

Ferreira ML, Barbosa MF, Gomes EPC, do Nascimento APB, de Luca EF, da Silva KG et al (2021) Ecological implications of twentieth century reforestation programs for the urban forests of São Paulo, Brazil: a study based on litterfall and nutrient cycling. Ecol Process 10(1):1–13

Fontão PAB, Ferreira RMA (2022) As Chuvas no Sistema Cantareira: Avaliação dos Relexos no Manancial Visando a Seguranca Hídrica da Região. Metropolitana de São Paulo. Revista de Geografia-PPGEO-UFJF 12(2):218–238

Gerberg J (2015) A megacity without water: São Paulo's drought. Time

Gozzo LF, Palma DS, Custodio MS, Machado JP (2019) Climatology and trend of severe drought events in the state of São Paulo, Brazil, during the 20th century. Atmos 10(4):190

Haddad EA, Teixeira E (2015) Economic impacts of natural disasters in megacities: the case of floods in São Paulo, Brazil. Habitat Int 45:106–113

James PE (1933) The surface configuration of southeastern Brazil. Ann Assoc Am Geogr 23(3):165–193

Kaźmierczak A, Cavan G (2011) Surface water flooding risk to urban communities: analysis of vulnerability, hazard and exposure. Landsc Urban Plan 103(2):185–197

Liebmann B, Jones C, de Carvalho LM (2001) Interannual variability of daily extreme precipitation events in the state of São Paulo, Brazil. J Clim 14(2):208–218

Millington N (2021) Stormwater politics: flooding, infrastructure, and urban political ecology in São Paulo, Brazil. Water Altern 14(3):866–885

Municipality of São Paulo (2021) Parques Lineares. Man. Desenho Urbano e Obras Viárias. URL https://manualurbano.prefeitura.sp.gov.br/manual/6-infraestrutura-verde-e-azul/6–2-infraestrutura-verde-e-azul/6–2-1-parques-lineares

Neves FDC (2001) Getúlio e a seca: políticas emergenciais na era Vargas. Revista Brasileira de História 21:107–129

Peixoto-Mehrtens C (2010) Urban space and national identity in early twentieth century São Paulo, Brazil: crafting modernity. Palgrave Macmillan, New York

Pessoa DF (2019) The Tietê River straightening process and its implications in the City of São Paulo, Brazil. Paisag Ambiente: Ensaios, São Paulo 30(44):e158617

Post GC, Chang H, Banis D (2022) The spatial relationship between patterns of disappeared streams and residential development in Portland, Oregon, USA. J Maps 18(2):210–218

Quintslr S, Peregrina Puga B, Octavianti T (2021) Mobilization of bias: learning from drought and flood crises in São Paulo, Rio de Janeiro and Jakarta. Water Int 46(6):861–882

São Paulo City Executive Secretary of Communication (2017) Maior piscinão da cidade de São Paulo é inaugurado na Zona Leste. https://www.nossasaopaulo.org.br/2017/02/02/maior-piscinao-da-cidade-de-sao-paulo-e-inaugurado-na-zona-leste/

Sheffield J, Andreadis KM, Wood EF, Lettenmaier DP (2009) Global and continental drought in the second half of the twentieth century: severity–area–duration analysis and temporal variability of large-scale events. J Clim 22(8):1962–1981

SOSMA/INPE – SOS Mata Atlântica/Instituto Nacional de Pesquisas Espaciais (2018) Atlas dos Remanescentes Florestais da Mata Atlântica Período 2016–2017. Technical Re- port, Arcplan, São Paulo, p 63

Tomás LR, Soares GG, Jorge AAS, Mendes JF, Freitas VLS, Santos LBL (2022) Flood risk map from hydrological and mobility data: a case study in São Paulo (Brazil). Trans GIS 26(5):2341–2365

Wikipedia (2023) https://en.wikipedia.org/wiki/Tiet%C3%AA_River#/media/File:Rio_tiete.jpg

Young AF, Papini JAJ (2020) How can scenarios on flood disaster risk support urban response? A case study in Campinas metropolitan area (São Paulo, Brazil). Sustain Cities Soc 61:102253

# Chapter 11
# Mexico City, Mexico

**Abstract** This chapter discusses the history of the hydraulic infrastructure that created the conditions for the modern city, as well as the trade-offs involved in producing the contemporary urban water phenomenon. Subject to increasing threats of flooding, along with drought intensity, Mexico City must contend with the intensification of the water cycle in the context of interannual variability, making predictions somewhat more reliable (especially for droughts), while solutions remain complicated and piecemeal. Complications in water services to the densely populated periphery are combined with water quality crises in the city, as water drawn from aquifers contributes to subsidence, which adds to pipe breakages and leaks. The subsidence adds to the imperviousness of an already developed surface, making pluvial nuisance flooding more acute. In recent years, however, some intrepid architects, designers, and ecologists have presented novel solutions of integrated and holistic water management, developing "water squares" that allow infiltration and filtration of surface runoff, adding the possibility of recycled grey water.

**Keywords** Infrastructure · Desagüe · Subsidence · Water squares · Integrated water management

## 11.1 Introduction

When one thinks of the image of Mexico City, the city is crowded but rich in historical buildings, being the oldest and largest capital city in the Americas, founded by the indigenous people. Today, as the home of over 21 million people, Mexico City, while producing the highest gross domestic product, faces myriad water challenges, including clean water supply, floods, and droughts. Originally developed on the lakebed in the high plateaus in the Mexico Valley, the city expanded to surrounding volcanic slopes over time (see Fig. 11.1).

The elevation of the Mexico basin, which surrounds the city, ranges from 2000 to over 5000 meters. The topsoils in the valley are rich in clay and silt, which function as conduits of water during storm events as clays expand quickly. The unique



H. Chang, A. R. Ross, *Climate Change, Urbanization, and Water Resources*, https://doi.org/10.1007/978-3-031-49631-8_11

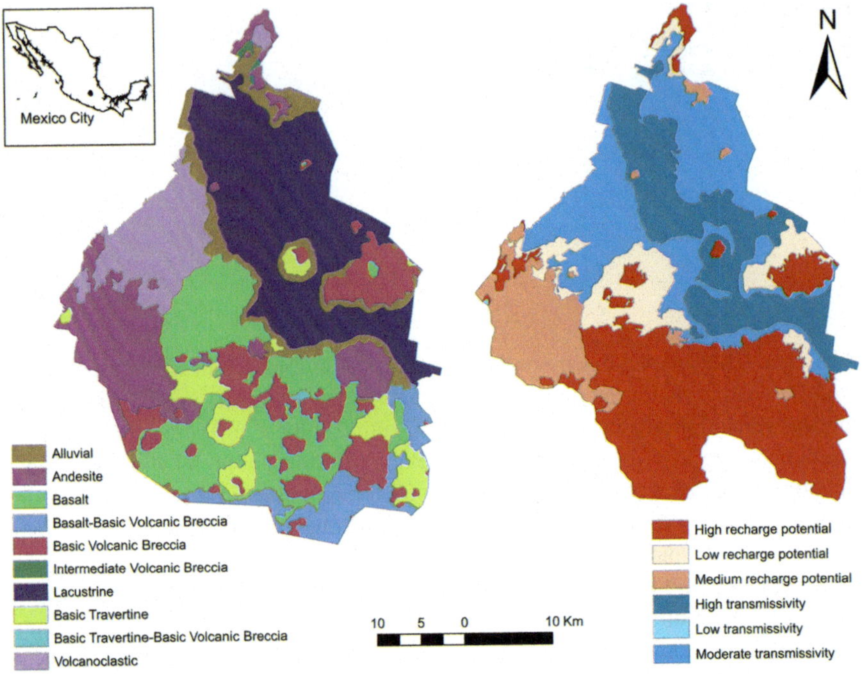


**Fig. 11.1** Location of Iztapalapa district and its geology (left) and hydrogeology (right) (map layer sources provided by INEGI 2023)

combination of topography and geology has regulated water flows historically. However, the human encroachment of natural areas, once occupied with seasonal water, has made its residents vulnerable to water-related hazards such as land subsidence, floods, and droughts (Burns 2009). The disappearance of seasonal playa lakes, which once controlled the flow of water during the wet season and provided water during the dry season, has exacerbated the water-related problems in the region (Alcocer and Bernal-Brooks 2010). Today, there are several large lakes. Unfortunately, these water bodies are largely contaminated, unsuitable for human consumption. With the rise in air temperature and increasing precipitation variability, the city now faces more challenges in managing its water resources.

## 11.2 Climate and Geography

Like all megacities around the world, Mexico City drew migrants from rural areas during much of the twentieth century. As shown in Fig. 11.2, while the population in the urban core has declined continuously since the 1970s, the population in the balance of the federal district and outside of the federal district increased exponentially since 1950, reaching over 92% of the total population in the Mexico City

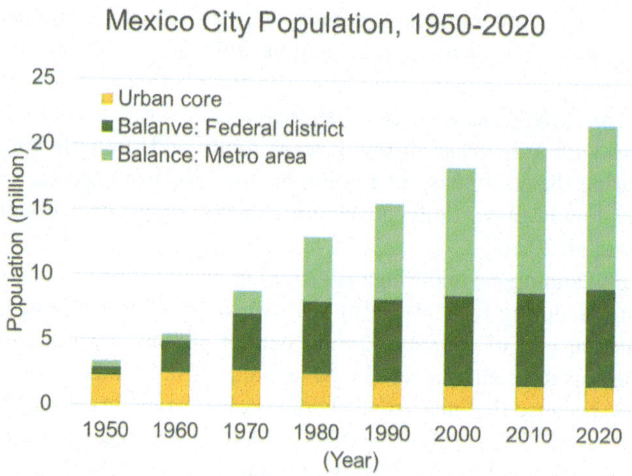


**Fig. 11.2** Population change in Mexico City urban area, 1950–2020. (Data sources: Census and United Nations available from INEGI (2023) and Macrotrends (2023). Figure design adapted from Cox 2019)

metropolitan area. To accommodate the growing population, the city expanded to the outskirts of the city by reclaiming lakebed and volcanic rocks (highly permeable basalt and andesite rock) that once held water during the wet period and released water gradually during the dry period. Additionally, the mixed forests of diverse species in surrounding hilly areas of the city, which once functioned as a buffer against strong winds and the erosion of topsoils and regulated the amount of water that goes into the soil and air, disappeared as a result of sprawl urban development. The disappearance of forests has increased surface runoff and thus flooding in more recent decades. With a high population density of 6000 people/km$^2$, per capita water availability (74 m$^3$/year) is lowest in the country.

Located at high altitudes, the city has high a subtropical highland climate, offering mild climates throughout the year. The mean annual temperature is around 15 °C, and summer temperature rarely reaches above 30 °C. The city has distinct summer wet (May–October) and dry winter (November–April) seasons with nearly 95% (780 mm) of annual precipitation falling during the wet season. Some 65% of annual variability in the Mexican climate results from the teleconnections comprising the El Niño-Southern Oscillation (ENSO). During the warm El Niño phase, the Walker circulation weakens substantially, so the normal easterly winds that converge at the equator calm down and even reverse course. Consequently, the warm surface waters of the Pacific that would be sheared away from the western coast of the Americas by those winds remain stagnant, leaving the surface of the eastern Pacific to heat up considerably. The combination of the eastern Pacific heating and the reversal of easterly trade winds causes convectional uplift that feeds large storm clouds, which follow a course into Latin America, from Peru to the western coast of Mexico.

Yet, the impact of interannual variability on Mexico City is complicated by its continentality, as well as its topography and the influence of the Atlantic Ocean. On the other side of the continent, the climate in the Caribbean dries up during the warm phase of El Niño as a result of an increase in vertical wind shear over the region, along with a decrease in low-level vorticity and strengthening of column stability. During the cold phase of ENSO, by contrast, wind speeds even out with altitude, and low-level vorticity advection can contribute to cyclonic formation (Dominguez et al. 2021).

A study by Dominguez and Magaña (2018) indicated a lower than median average precipitation during El Niño warm phases, with rainfall declining especially in the southwestern part of the city. During neutral phases, precipitation is seen to increase, whereas the cool phase of La Niña indicates no change. This phenomenon likely occurs, because the inter-tropical convergence zone thrusts southerly during the warm phase, while easterly wave tracks die down. At the same time, the interplay of a cold Pacific Decadal Oscillation and a cold ENSO can send tropical cyclones into southwestern Mexico without increasing total cyclonic activity, while the Atlantic Multi-decadal Oscillation has a complementary impact on ENSO cycles in the Caribbean (Dominguez et al. 2021).

## 11.3 Urban Development

As a result of the uneven distribution of rainfall throughout the year, the city has experienced chronic droughts and floods. Historically, Mexico City had 136 droughts between 1450 and 1900 (Mendoza et al. 2005). Floods are the most common water-related hazard in Mexico City. According to desinventar.net, there were 687 cases of flood reports with a loss of $1878.1 million damage between 1970 and 2013. According to the International Institute for Environment and Development Research Report (2007), the city is already unable to cope with climate-induced water-related hazards.

Archeological evidence exists of indigenous hydrologic infrastructure combating floods as early as a 100 years prior to contact with Europeans. Underlying the modern site of Mexico City are the ruins of the Aztec empire's military seat, Tenochtitlan, a place that experienced the same problems of flooding and accompanying issues. Constructed two centuries before the arrival of Spanish conquistadors in 1519, Tenochtitlan sat on an island within a lake system within the Valley of Mexico (Tellman et al. 2018). Mountain rivers fed into Lake Chalco, which then poured into Lake Xochimilco, while Lake Texoco received water from those sources and the river-fed Lake Mexico, which in turn fed back into Toxoco. With its level raised just two meters above Lake Texoco surrounding it, Tenochtitlan faced frequent and debilitating flooding (Torres-Alves and Morales-Nápoles 2020).

To combat the floods, the Aztecs built dikes, carved out canals, constructed aqueducts, dug irrigation ditches and drainage systems including ponds and pools, and developed *chinampas*—floating gardens that made use of the prevalent water for agriculture. Some of these projects reflected extensive engineering, for instance, the Nezahualcoyotl dike, built in the 1440s, extended 16 km at 8 m in height and built with a width of 3.5 m. The dike utilized entrenched posts underlying a base of sand and stones into which rows of wooden logs could be stacked up to form a kind of hollow container filled in with more rocks and sand (Tores-Alves and Morales-Nápoles 2020).

No evidence suggests that, during its decades of existence, the dike failed or overflowed (Tores-Alves and Morales-Nápoles 2020). After the Spanish conquest, however, it was damaged and destroyed, followed by frequent flooding over the course of the next centuries. Spanish practices of husbandry led to erosion and the continued collapse of Aztec infrastructure, and massive floods tore through the basin every 25 years—in 1555, 1580, and 1604—before the reconstructed capitol of Mexico City finally resorted to complete drainage of the entire lake (López).

The idea of desagüe, or dewatering, emerged in the early seventeenth century through the efforts of Enrico Martínez, a German-born engineer who advanced the idea of a hand-dug canal into the Gulf of Mexico. By contrast, the Dutch hydraulic engineer Adrian Boot sought to develop similar hydraulic systems to Holland, making use of the lacustrine systems rather than dispatching them. While the city ultimately committed to Martínez's design, Boot was correct that the drainage of lakes would not prevent flooding but could ultimately impose problems of water quantity. The initial desagüe failed, however, as the Cuauititlán River continued to capture precipitation, causing chain overflows throughout the lake system (López).

Further desagüe projects rose and fell with flood risk perception, increasing after a terrible flood in 1629 and again during the 1760s. The massive undertakings involved the labor of tens of thousands of indigenous people at a time, siphoning the workforce of local townships while depleting the hydrological resources necessary for their social and economic reproduction. Throughout the eighteenth century, the desagüe continued to divert traditional ecosystem value from indigenous residents of the basin, breaking up transportation networks, desiccating important croplands, and destabilizing traditional uses (Tellman et al. 2018). As a result of the ecological transformations and economic disturbances, riots ensued along with everyday practices of resistance (Candiani 2012). Regardless, vital agricultural production was dismantled, and soils were irretrievably damaged, resulting in a worsening of seismic activity.

Persistent flooding continued, along with the persistent problem of water deficit, into the late nineteenth century, at which point, Mexico City brought together a consolidated, urban water management system in keeping with the overall movement of modernization. Renewed efforts to tunnel water out of riverbeds and replace them with roads were coupled, several decades later, with the Lerma and Cutzamala systems (in the 1950s and 1970s, respectively) of reservoirs, pumping stations, and treatment plants (Tellman et al. 2018).

## 11.4 Drinking Water Issues Associated with Urban Development

About two-thirds of Mexico City obtains its drinking water from the underground aquifer of the Mexico City metropolitan zone (Zona Metropolitana de la Ciudad de México—ZMCM). The extracted water is distributed via a complex network of pipes within the metropolitan area. Not everyone has access to tap water and even if it does, water supply is intermittent, heavily relying on trucked water.

Developed between the 1940s and 1980s, the Lerma system transfers water from the upper Lerma basin's groundwater aquifers to the city. Initially, the water was drawn from streams, but with the decline of surface water availability, groundwater extraction started in 1951, which coincided with rapid population growth. Today, the Lerma water system only supplies about 6% of the total supply; the rest goes to the industrial and commercial agricultural sectors. The other system is the Cutzamala, which together with the Lerma supplies almost 30% of water to the city (Carrera-Hernandez 2018; Freeman 2019). Developed in 1982, the Cutzamala system collects water in seven dams in the Cutzamala River. The water is transferred to Mexico City and the State of Mexico via a 127 km aqueduct using natural elevation difference. The system is particularly vulnerable to long-term drought because the reservoirs only store 1.2 years of historical water supply to the Mexico City metropolitan area (Freeman et al. 2020). Yet, despite its frequent rains and the original abundance of water, Mexico City today has the lowest water availability in Mexico per capita (Chen and Bilton 2022).

Mexico City grew tremendously during the twentieth century, from around one million in 1900 to more than nine million in 1970, and by 1970, the percentage of its population that lived in informal settlements increased from 14 to 50 (Alba and Potter 1986; Cruz 1991). This rapid growth began to taper off in the 1980s, following the catastrophic 1985 earthquake that destroyed much of the urban area, causing an exodus to peripheral areas and encroaching on agriculturally and ecologically important areas (Lerner et al. 2018).

In the wake of the disaster, an unfolding economic crisis extending into the 1990s cut short infrastructure development. Using river water for drinking is not feasible since most surface rivers or streams have been converted into stormwater or sewage drains (Martinez et al. 2015).

As a result, informal solutions, such as the *pipeo* system, which brought water out to informal settlements by truck, increasingly became patches to plug the holes in water supply. Proximity to truck routes now determines the accessibility of water supply to many in informal settlements. In addition, sewer systems are often inadequate in informal settlements, leading to systemic health problems (De Alba 2017).

Water quality in Mexico City combines with quantity to create problems throughout the hydrologic system. In 2014, just 71% of the city's water could obtain the standard of acceptable potable water, and 15% of the total provision came from overdraft of aquifers (De Alba 2017). This overdraft causes the land surface to sink down between 5–36 cm per year, which in turn leads to structural damage to buildings and pipes—a feedback problem of water inefficiency in a city that loses around 40% of water to leakage (Chen and Bilton 2022). Informal settlements suffer the most from the water stress, as the land subsidence conspires with inefficiency and poorly planned state subsidies to disenfranchise large portions of the city's residents (De Alba 2017; Chen and Bilton 2022).

There is substantial spatial variation in residential water demand in Mexico City. Water use is tightly coupled with the socio-spatial structure of the city. According to the study by Ramos-Bueno et al. (2021), high residential water demand in areas coincides with high social development index and houses with metering devices, while dwelling density is negatively associated with water demand. Local hotspot analysis clearly shows distinct clustered spatial patterns in these water demand, infrastructure, and explanatory variables (Fig. 11.3). This finding suggests that

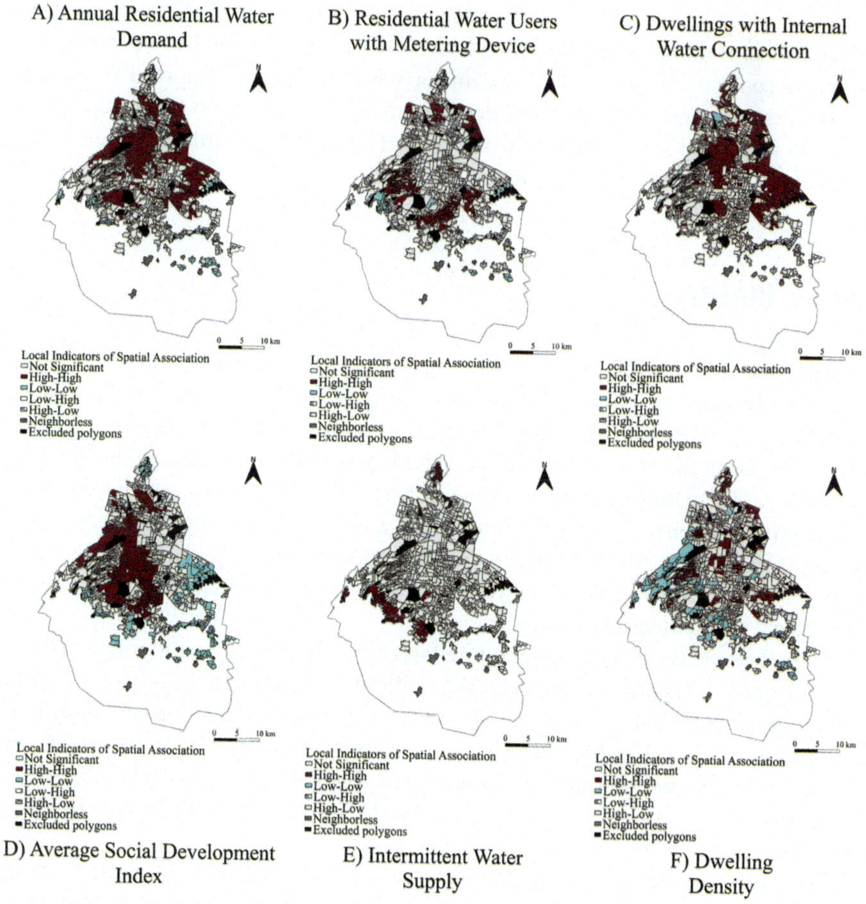


**Fig. 11.3** Local indicators of spatial autocorrelation (LISA) for the dependent and independent variables for the analysis of annual residential water demand in Mexico City neighborhoods (clusters of high values are in dark red and clusters of low values are in light blue). (Source: Ramos-Bueno et al. 2021)

spatial patterns of water infrastructure and density of development play important roles in water demand in Mexico City. Inevitably, water infrastructure is not evenly distributed over space. Typically, densely developed areas are occupied by people who do not have the means to obtain clean drinking water (Satterthwaite et al. 2007).

An attempt has been made to improve drinking water security. For example, in 2014, Mexico City initiated the Water Sustainability Law (Ley de Sustentabilidad Hídrica) that requires "the city government to conform to the principles of human rights, sustainability, transparency, and shared responsibility" (Meehan 2019). While this city law is pioneering in that it provides nationally unavailable regulatory and enforcive measures for improving basic human rights to access to safe drinking water and sanitation, it also generated some controversial issues, including a potential market for harvested or purified rainwater, privatization of water provision, not supplying piped water services to irregular settlements, and suspension of water services to late water bill payers (Meehan 2019). In response to the potential privatization of drinking water, Coordinadora Nacional Agua Para Tod@s, which is a national coalition of water activists, came up with the Citizen's National Water Law: "a set of governance principles and decision-making structures to realize the human right to water and sanitation in Mexico" (Meehan 2019). According to them, water should remain in the public domain.

## 11.5 Floods

During the urbanization process, extreme events got more frequent and intense. Ongoing urbanization resulted in an increase of 0.15 °C in annual minimum temperatures, leading to an urban heat island effect with a thermal gradient of more than 10 °C between the city center and the suburban region. According to the study by Ochoa, the frequency of extreme rain events (>20 mm/h) has also increased, while the timing of intense events shifted to earlier afternoon due to the urban heat island effect (Ochoa et al. 2015). In effect, the city heats up more quickly, leading to more rapid evaporation and ensuing convectional uplift, which provides the water vapor necessary for condensation and rain.

Together with climate change, future land cover change (forest to urban and forest to croplands) is projected to decrease infiltration capacity of the city by 8%, with urban land use changes causing the most substantial reduction in infiltration (Zambrano et al. 2017). By converting permeable surfaces that allow for groundwater recharge into non-permeable surfaces that channel runoff into low-lying areas, land use changes are projected to contribute to an increase in flood risk by up to 10% (Zambrano et al. 2017).

According to a study by Zambrano et al. (2018), the risk of small-scale flooding extends to more than half of Mexico City, while nearly a quarter of the city risks flooding on a large scale. Much of the most exposed area of Mexico City lies in the "peri-urban interface," where informal settlements extend around the periphery and socio-natural factors such as waste contribute to flooding intensity. In a single event,

Hurricane Dean hit Mexico in 2008 with 79 mm of rain in 3 h, straining the sewers and unleashing torrents of mud, debris, and waste (Lankao 2010).

Flooding is not only caused by immediate precipitation but developmental problems as well. On June 1, 2000, for instance, the southeastern peri-urban interface faced a massive inundation when heavy rains contributed to the breach of a canal holding back an open sewage canal—a disaster traceable not only to the chronic flooding but also to the socio-historical problem of development which left vast areas disconnected from sewer systems, with the only alternative being open sewer channels (Aragon-Durand 2007).

To confront the problem of flooding, city administrators focus on clearing the waste from sewers that blocks the rapid drainage of heavy rains. By removing blockage and pumping water into sewers, administrators hope to confront a major contributing factor to floods, but Zambrano's study shows that the approach remedied only small floods (2018). Instead of simply clearing waste from sewers, Bonasia et al. (2023) show that the city should focus on stormwater retention in the abandoned tracts prevalent throughout the city, which often comprise undefined soils in non-urbanized areas with low permeability. While runoff from these tracts tends to contribute to flooding, developing ponds and other hydrologic infrastructure would help the rainwater infiltrate into aquifers and decrease runoff potential.

## 11.6 Droughts

While floods often feature in the conversation on water issues in Mexico City, droughts are also incredibly important. Given the multiple, intertwined water problems and risks, it becomes difficult to prioritize the most pressing need. While land-use actors generally see urbanization as the greatest risk, water-use actors view flooding and health as far more urgent. Both actors agree, however, that scarcity is a major threat (Lerner et al. 2018).

Nine serious drought periods occurred from the year 1450 to 1900, with strong and very strong droughts often coinciding with ENSO's warm period (Mendoza and Ortiz Guitart 2008). These droughts tended to interfere with normal agricultural life, leading refugees to flee the desiccating fields to population centers where infrastructure could not sustain them. According to tree-ring chronologies and crop yields, paired with medical data, it is clear that these historic droughts led to 19 of 22 typhus epidemics between 1655 and 1918 (Burns et al. 2014).

Six more droughts occurred between 1948 and 2006, causing huge amounts of economic damage. In 2022, an extreme drought ravaged two-thirds of Mexico, drawing down the Cutzamala System to its lowest point since 1996. While the state of Michoacan broke ground on a new canal to draw water from Cutzamala into its own water-stressed region, the Mexican air force puts forward efforts at "cloud seeding," releasing silver iodine into clouds over the reservoirs to encourage droplet formation and subsequent rainfall (O'Boyle et al. 2023).

The changes in temperature and extreme precipitation events induced by climate change could be worsened under global warming scenarios. Climate change projections in the Mexico City area forebode a dire future. The mean annual temperature is projected to increase between 1.3 and 1.9 °C by the mid-twenty-first century. Summer months' temperatures are even projected to be higher than 2 °C, leading to an increase in evapotranspiration. Additionally, the mean annual rainfall is projected to decrease in all scenarios considered. Such changes in precipitation and temperature are likely to result in more frequent seasonal droughts. While there is an uncertainty in projecting future flow, a decrease of 10–17% annual water availability is likely (Martinez et al. 2015).

## 11.7 Sustainable Alternatives

Mexico City's administration tends to support more top-down solutions to water issues, relegating decentralized planning to secondary status. To be sure, the sunken costs involved in the city's heavy reliance on a hydrologic regime implemented during the seventeenth century and fully realized in the nineteenth century requires a kind of approach more oriented to solving the continuing problems of past decisions than to devising a new and innovative regime. To achieve water resource resilience at the urban scale, one needs to have coordinated efforts that encompass the different interests of various stakeholders (Eakin 2022; Eakin et al. 2016).

In Mexico City, many informal institutions exist, and without active engagement of these institutions, the current gaps in the decision-making process would not be narrowed, making it impossible to achieve reduce water-related vulnerability (Lerner et al. 2018). Given that infrastructure planning and decisions have long-lasting effects on future hydrologic risks, decisions need to be made right in the first place. However, novel and transformative developments are increasingly turning the city's planning apparatus toward more sustainable ideas that, through persistent efforts, could provide radical new ways of doing urban resilience.

One example of surprise solutions lies in the farmlands outside the city. In spite of the frequent floods, the drainage system's mixture of waste and runoff makes much of the water unusable, pumping it outside of the valley to the fields of the Mezquital Valley in the north. Filled with plastics, parasites, and bacteria, this "black water" irrigates cropland dedicated to animal feed, allowing the water to percolate through the cleansing earth and gather in a more filtered aquifer below (Delgado-Ramos 2015). While problematic in many respects, the case of Mexico's "black water" filtration offers an example of connecting the hydrosphere to the lithosphere and remedying, in part, water quality issues (although bioremediation on soils will become increasingly important).

The generation of "black water" indicates problematic water treatment, and scientists argue that Mexico City will need more comprehensive approaches than slapdash solutions to resolve the key problems at work in its water dilemma. Freeman et al. (2020) conducted a comprehensive study of Mexico City's water system resilience under a range of precipitation and temperature change scenarios. By engaging 34 private and public institutions, academic organizations, and citizen groups over a two-year period, their collaborative workshops produced some of the most comprehensive and interesting potential adaptive strategies under climate uncertainty. They first identified the system's vulnerability to climate change by 2050 and explored possible options to meet the water demand. Their study shows that urban areas such as Mexico City should be seen more broadly as one part of larger freshwater systems, adding that socio-ecological efforts based on participatory planning and demonstrated performance will be most effective in dealing with the city's water issues comprehensively (Freeman et al. 2020).

In 2007, Mexico City's administration attempted to draft a comprehensive *Plan Verde de la Ciudad de Mexico*, or the "Green Plan." As problems of sustainability already came to a head in the first decade of the twenty-first century, Mexico's planners hoped to engage in a participatory process with the public but were criticized from the beginning for failing to consult with a substantial proportion of the population. Since the plan came from the Ministry of the Environment and not the whole government, it also faced concerns of "horizontal fragmentation" within the government. Also, the participation of NGOs garnered critiques from political groups who viewed the process as elitist and bureaucratic rather than grassroots. While the next few years indicated some successes for the Green Plan—increased bicycle use and decreased car emissions, for instance—the plan dwindled under the ensuing city administration under Miguel Ángel Mancera (Madero and Morris 2016).

In lieu of an ongoing, clear plan that can systematize a concerted push to confront climate change, advances are still being made in a number of ways. Service-providing units of green infrastructure in Mexico City have been shown to add ecosystem services such as flood control and runoff mitigation, indicating the possibility of an increase in decentralized urban improvement (Calderon-Contreras and Quiroz-Rosas 2017). One study of the neighborhood of Valle del Chalco showed reason for optimism in the development of resilient infrastructure for water management but also detected some lack of consistency in the political process and policy-forming entities with regard to the values underlying resilience itself (Mahajan et al. 2022).

The problem of impervious surfaces contributing to flooding is also being confronted through fascinating approaches. First launched in Rotterdam in 2013 as part of the Water City 2035 program, "Water Squares" are public squares or plazas with water capture infrastructure. Recessed into the ground and often connected to the aquifer through infiltration devices, rain gardens, and pools, the squares can add green space by introducing native plants with long and dense root networks while also creating interesting and artistic design features in public space. Kept dry until intense rain events, the Water Square channels area pluvial floods into an intentional drainage system that recharges aquifers while removing water from the city streets. Thus, instead of attempting to channelize excess water to the periphery, these newer features can pull water through the city quickly, increasing resident safety while counteracting the problem of urbanized impermeability (Ilgen et al. 2019).

Mexico City's densely populated, low-lying Iztapalapa district, which lies within former Lake Texacoco, will serve as the site for the pilot Water Square—Parque Hídrico Quebradora. With support and input from the community, Universidad Nacional Autónoma de México architects created a 3.8-ha site to be energy self-sufficient and planted with local, endemic vegetation. Capable of harvesting rainwater while treating wastewater for use in public toilets, Parque Hídrico Quebradora comprises terraced plazas making use of a natural slope, with some spaces dedicated to recreational facilities such as covered basketball courts and others to rain fountains, wetlands, and flood pools (Cohen and Mancera 2018).

Although the Parque Hídrico Quebradora manifests a grander design than most "Water Squares," which are often intended as smaller, efficient interventions scattered throughout the more vulnerable parts of the city, it presented the first "water smart work" in Mexico (Vargas Lara 2018). According to architect Lora Castro Reguera, the intention behind the gold-medal winning project was "to transform the way people relate to water," adding that more architects need to consider "developing interventions that vertically integrate different kinds of solutions, taking into account infrastructure, culture, urbanism, the environment, and even the economy" (Flaksman et al. 2021).

## 11.8 Conclusions

While future hydroclimate scenarios look gloomy for Mexico City, the impacts of water-related hazards depend on how the city can increase its adaptive capacity in the future. Mexico's efforts to resolve some of the city's most pressing water issues have borne some fruit in recent years, but redoubled efforts are needed to continue the momentum. Yet, political changes tend to throw off scheduled projects and agendas. As a megacity, Mexico City can set an example by creating a participatory climate agenda to remain constant regardless of the political tendencies that gain or lose power.

Climate change threatens to increase storm intensity in Mexico City, while also lengthening and deepening periods of drought. Mexico's unique development history only intensifies those problems. Created in a low-lying lake system drained over a period of hundreds of years, Mexico City's introduction of impervious surfaces causes decreased infiltration and greater. As the megacity grows, it relies on well water, drawing down the aquifer, increasing contamination, and causing subsidence that worsens the problem of imperviousness (Mazari and Mackay 1993). During drought periods, the dilemma of water quantity increases, even as poorer districts around the city lack basic water services, including sanitation and piped water.

The city's Green Plan shows that participatory projects on a large scale can work, but administrators need to harness the robust grassroots social infrastructure of democratic practices deployed in poorer areas in order to implement such efforts in ways that have staying power. Parque Hídrico Quebradora provides an excellent example of an in situ participatory project that integrates community use with multiple water management services, including water quality, reuse, and flood control. At the same time, it is only one large structure whose provisions will be limited to the general area. To truly operationalize the utility of Water Squares, such efforts can be dispersed throughout the city in smaller form but on a larger scale.

To curate an adaptive strategy for climate hazards, Mexico can continue to build trust with local communities by tapping into the deep well of social organization and engagement. Decentralized power can work in the city's favor by dispersing climate projects throughout the city instead of concentrating on one large site at a time. Areas with greater vulnerability to extreme hazards tend to have less input into policy, and without a strong regulatory agenda and the agencies to enforce it, the sovereignty of the people is weakened.

## References

Alba F, Potter JE (1986) Population and development in Mexico since 1940: an interpretation. Popul Dev Rev 12(1):47-10.2307/1973351

Alcocer J, Bernal-Brooks FW (2010) Limnology in Mexico. Hydrobiologia 644(1):15–68. https://doi.org/10.1007/s10750-010-0211-1

Aragón-Durand F (2007) Urbanisation and flood vulnerability in the peri-urban interface of Mexico City. Disasters 31(4):477–494. https://doi.org/10.1111/disa.2007.31.issue-4, https://doi.org/10.1111/j.1467-7717.2007.01020.x

Bonasia R, Borselli L, Madonia P (2023) Analysis of flow and land use on the hydraulic structure of Southeast Mexico City: implications on flood and runoff. Land 12(6):1120. https://doi.org/10.3390/land12061120

Burns E (2009) Repensar la cuenca: la gestión de ciclos del agua en el Valle de México. Universidad Autónoma Metropolitana de México, México

Burns JN, Acuna-Soto R, Stahle DW (2014) Drought and epidemic typhus Central Mexico 1655–1918. Emerg Infect Dis 20(3):442–447. https://doi.org/10.3201/eid2003.131366

Candiani V (2012) The Desagüe reconsidered: environmental dimensions of class conflict in colonial Mexico Abstract. Hisp Am Hist Rev 92(1):5–39. https://doi.org/10.1215/00182168-1470959

Calderón-Contreras R, Quiroz-Rosas LE (2017) Analysing scale quality and diversity of green infrastructure and the provision of urban ecosystem services: a case from Mexico City. Ecosyst Serv 23:127–137. https://doi.org/10.1016/j.ecoser.2016.12.004

Carrera-Hernandez JJ (2018) A tale of Mexico's most exploited—and connected—watersheds: the basin of Mexico and the Lerma-Chapala Basin. WIREs Water 5(1). https://doi.org/10.1002/wat2.2018.5.issue-1, https://doi.org/10.1002/wat2.1247

Chen Y, Bilton AM (2022) Water stress Peri-urbanization and community-based water systems: a reflective commentary on the metropolitan area of Mexico City. Frontiers in Sustainable Cities 4. https://doi.org/10.3389/frsc.2022.790633

Cohen MP, Mancera LCR (2018) Design with water. ReVista (Cambridge) Spring 17(3):72–75,82

Cox W (2019) Expanding, Productive Mexico City: the evolving urban form. https://www.newgeography.com/content/006465-expanding-productive-mexico-city-the-evolving-urban-form

De Alba F (2017) Challenging state modernity: governmental adaptation and informal water politics in Mexico City. Curr Sociol 65(2):182–194. https://doi.org/10.1177/0011392116657288

Delgado-Ramos G (2015) Water and the political ecology of urban metabolism: the case of Mexico City. Journal of Political Ecology 22(1). https://doi.org/10.2458/v22i1.21080

Dominguez C, Magaña V (2018) The role of tropical cyclones in precipitation over the tropical and subtropical North America. Front Earth Sci 6. https://doi.org/10.3389/feart.2018.00019
Dominguez C, Jaramillo A, Cuéllar P (2021) Are the socioeconomic impacts associated with tropical cyclones in Mexico exacerbated by local vulnerability and ENSO conditions? Abstract International Journal of Climatology 41(S1). https://doi.org/10.1002/joc.v41.S1, https://doi.org/10.1002/joc.6927
Eakin H (2022) Weathering risk in rural Mexico: climatic, institutional, and economic change. University of Arizona Press, Tucson
Eakin H, Lerner AM, Manuel-Navarrete D, Aguilar BH, Martínez-Canedo A, Tellman B et al (2016) Adapting to risk and perpetuating poverty: household's strategies for managing flood risk and water scarcity in Mexico City. Environ Sci Pol 66:324–333
Flaksman C, Wesseler S, Castro-Reguera L, Ambrosi JP (2021) The image of water. The Architectural League NY, August 10. https://archleague.org/article/the-image-of-water/
Freeman S (2019) The baffling predicament of Mexico City's water: part two. Pacific Council on International Policy, October 8:2019. https://www.pacificcouncil.org/newsroom/baffling-predicament-mexico-city%E2%80%99s-water-part-two
Freeman SSG, Brown C, Cañada H, Martinez V, Nava AP, Ray P, Rodriguez D, Romo A, Tracy J, Vázquez E, Wi S (2020) Resilience by design in Mexico City: a participatory human-hydrologic systems approach. Water Secur 9:100053
Ilgen S, Sengers F, Wardekker A (2019) City-To-City learning for urban resilience: the case of water squares in Rotterdam and Mexico City. Water 11(5):983. https://doi.org/10.3390/w11050983
INEGI (2023) Population total, Census, https://inegi.org.mx/default.html
Lankao PR (2010) Water in Mexico City: what will climate change bring to its history of water-related hazards and vulnerabilities? Environ Urban 22(1):157–178. https://doi.org/10.1177/0956247809362636
Lerner AM, Eakin HC, Tellman E, Bausch JC, Aguilar BH (2018) Governing the gaps in water governance and land-use planning in a megacity: the example of hydrological risk in Mexico City. Cities 83:61–70
Macrotrends (2023) Mexico City, Mexico Metro Area Population 1950–2023, United Nations - World Population Prospects, https://www.macrotrends.net/cities/21853/mexico-city/population
Madero V, Morris N (2016) Public participation mechanisms and sustainable policy-making: a case study analysis of Mexico City's plan Verde. J Environ Plan Manag 59(10):1728–1750. https://doi.org/10.1080/09640568.2015.1085841
Marzari M, Mackay DM (1993) Potential for groundwater contamination in Mexico City. Environ Sci Technol 27:794–802
Cruz SC (1991) Demography of Mexico City. The same problems with less population. Demos (Mexico City, Mexico) 4:23–24
Mahajan S, Chung M-K, Martinez J, Olaya Y, Helbing D, Chen L-J (2022) Translating citizen-generated air quality data into evidence for shaping policy. Abstract Humanities and Social Sciences Communications 9(1). https://doi.org/10.1057/s41599-022-01135-2
Martinez S, Escolero O, Perevochtchikova M (2015) A comprehensive approach for the assessment of shared aquifers: the case of Mexico City sustainable Water Resour Manag 1(2):111–123. https://doi.org/10.1007/s40899-015-0010-y
Meehan K (2019) Making space for practical authority: policy formalization and the right to water in Mexico. In: Sultana F, Loftus A (eds) Water politics: governance, justice and the right to water. Routledge, New York, pp 28, 13 p–41
Mendoza C, Ortiz Guitart A (2008) Spanish professionals in Mexico City: narratives on work and labour markets. Migracijske i etničke teme 24(4):323–340
Mendoza B, Jauregui E, Diaz Sandoval R, Garcia Acosta V, Velasco V, Cordero G et al (2005) Historical droughts in Central Mexico and their relation with El Niño. J Appl Meteorol 44:709–716

O'Boyle B, Romero H, Carrillo C (2023) Mexico City's thirst for water lays bare inequalities, changing climate. Reuters, April 27. https://www.reuters.com/world/americas/mexico-citys-thirst-water-lays-bare-inequalities-changing-climate-2023-04-27/

Ochoa CA, Quintanar AI, Raga GB, Baumgardner D (2015) Changes in intense precipitation events in Mexico City. J Hydrometeorol 16(4):1804–1820

Ramos-Bueno A, Perevochtchikova M, Chang H (2021) Socio-spatial analysis of residential water demand in Mexico City. Tecnología y Ciencias del Agua 12(2):59–110

Satterthwaite D, Huq S, Pelling M, Reid A, Romero Lankao P (2007) Building climate change resilience in urban areas and among urban populations in low- and middle-income nations. IIED research report commissioned by the Rockefeller Foundation, IIED, 112 pages

Tellman B, Bausch J, Eakin H, Anderies J, Mazari-Hiriart M, Manuel-Navarrete D, Redman C (2018) Adaptive pathways and coupled infrastructure: seven centuries of adaptation to water risk and the production of vulnerability in Mexico City. Ecol Soc 23(1):1

Torres-Alves GA, Morales-Nápoles O (2020) Reliability analysis of flood defenses: the case of the Nezahualcoyotl dike in the Aztec City of Tenochtitlan. Reliability Engineering & System Safety 203:107057. https://doi.org/10.1016/j.ress.2020.107057

Vargas Lara R (2018, October 4) First water smart work in Mexico: water park 'La Quebradora. IKI. https://ikialliance.mx/en/primera-obra-water-smart-en-mexico-parque-hidrico-la-quebradora/

Zambrano L, Pacheco-Muñoz R, Fernández T (2017) A spatial model for evaluating the vulnerability of water management in Mexico City Sao Paulo and Buenos Aires considering climate change. Anthropocene 17:1–12. https://doi.org/10.1016/j.ancene.2016.12.001

Zambrano L, Pacheco-Muñoz R, Fernández T (2018) Influence of solid waste and topography on urban floods: the case of Mexico City. Ambio 47:771–780

# Chapter 12
# Houston, United States of America

**Abstract** The City of Houston is considered the fossil fuel capital of the world, whose mid-century urban development strategy advanced the conditions of sprawl through car-friendly modernism. Always flood prone and vulnerable to hurricanes, Houston's disastrous experience of hurricane Harvey opened up the minds of its residents to the necessity of climate policies. The chapter examines different approaches taken by the city, from urban resilience strategies to storm mitigation through coastal wetland restoration. The model presented by Houston adds to the "water-sensitive" concept forwarded by Melbourne by including large-scale restoration policies of coastal storm breaks and habitat restoration that also benefit migratory corridors for birds and other wildlife. Houston's bold new direction is problematized by the continued economic reliance on fossil fuels, as well as the contradictory transportation policies of widening freeways at the expense of housing density. While the case of Houston introduces the considerable strategy of restoration as an important barrier to storm surges and the worst impacts of storm systems, its contradictions continue to reflect those of the broader modern world, maintaining the path of past policies shown to exacerbate climate change while implementing reforms that may confront symptoms of climate change without disturbing the primary causes.

**Keywords** Hurricanes · Wetland restoration · Climate resilience · Fossil fuels · Traffic

## 12.1 Introduction

On August 17, 2017, a hurricane turned the City of Houston, Texas, upside down. Causing damages of over $120 billion dollars, Hurricane Harvey reached Category 4 status as it made its way through the Gulf of Mexico, hitting coastal Texas before moving into Louisiana and then veering back over Houston. Bringing about the third 500-year flood in as many years, Harvey showed Houstonians and the world the new normal for climate systems in the subtropics—a deadly lesson that would



H. Chang, A. R. Ross, *Climate Change, Urbanization, and Water Resources*,
https://doi.org/10.1007/978-3-031-49631-8_12

forever change the lives of Houston's 2.3 million residents. The summer in 2017 had the highest ocean warming on record in the world and the Gulf of Mexico (Trenberth et al. 2018).

The climate diagnosis for Houston and, more generally, that part of the Gulf of Mexico indicates longer, hotter summers with a high probability of increased hurricane intensity (Nielsen-Gammon et al. 2020). Due to its flat topography, sprawling urban development, and massive fossil fuel infrastructure, these increased storms will pose enhanced hazards for not just low-lying residential districts but also all areas of the city, threatening housing as well as water quality and areal pollution spread by chemicals leaked from impacted oil and gas refineries.

Due to the increased hazards, Texans have increased efforts to tackle problems associated particularly with floods. Civilian groups supported by the municipal government have brought together local neighborhood community assemblies to deliberate and decide on the appropriate measures for meeting the challenges posed by climate-induced flooding brought on by the intensification of storms. While those local efforts encompass bridge restoration or elevation over bayous, sewer upgrades, and response networks, broader, regional proposals involve the implementation of wetlands restoration stretching from the coastal Gulf of Mexico into the Houston area in order to encourage the infiltration of storm waters into aquifers.

Due to the complex political character of the Houston area within the mostly conservative state of Texas, such climate-related policies can face major legislative obstacles. However, the exigency of increased flooding has become impossible to ignore for a state that once took center stage in the national discourse for challenging evolution in its state-sanctioned school textbooks (Armenta and Lane 2010). Where climate change has brought palpable and often fearsome effects, a tendency to shrug off its specter among the energy industry–connected elites of Houston has taken a back seat to the immediate need for adaptation.

## 12.2 Climate and Geography

No stranger to hurricanes, Houston lies at a subtropical latitude that renders its climate both warm and humid for much of the year. While there is a substantial interannual variability, Huston's annual precipitation and temperatures have increased over the past several decades (Figs. 12.1 and 12.2). Billed as a scenic, coastal property to settlers from the United States in the mid-nineteenth century, the Houston area was, in fact, largely a fetid swamp, where residents ran the significant risk of malaria from the region's prolific mosquito population (Speer Jr 1980). Fluctuations associated with climate change may cause diseases, such as dengue and chikungunya, to return (Chretien et al. 2015).

Like other states across the US South, Texas weather is always influenced by the cyclical patterns of El Niño. When the easterly trade winds in the central-south Pacific break down, the warm sea surface temperature that it typically pushes over to the Indonesian archipelago, causing a low-pressure zone of storm systems and

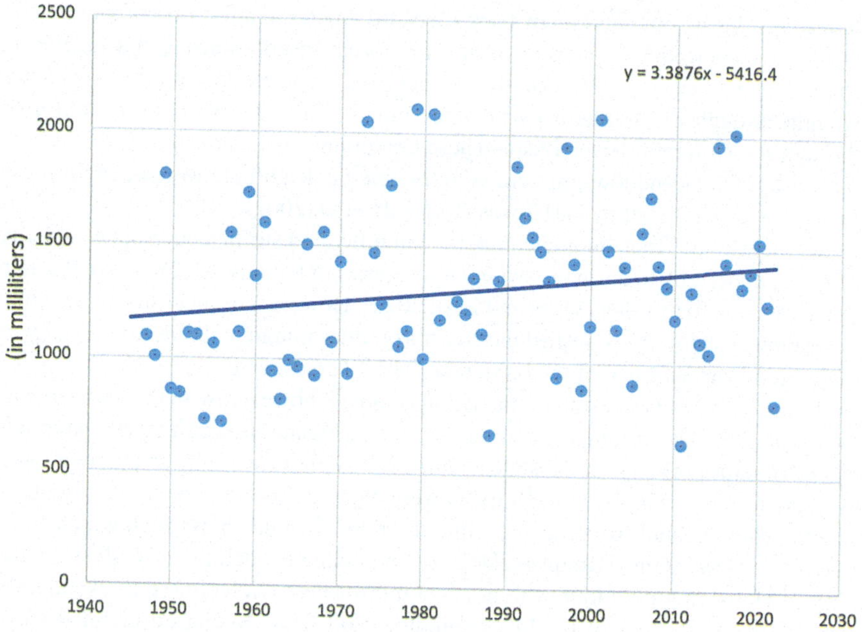


**Fig. 12.1** Annual precipitation at Hobby Airport, Texas. Note the outliers in 2011 (drought) and 2017 (Hurricane Harvey). (Source: Data from NOAA)

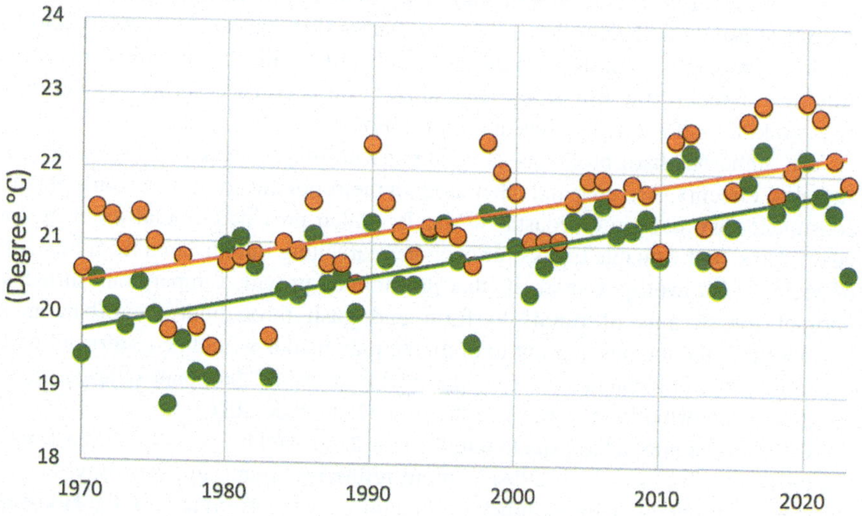


**Fig. 12.2** Annual temperature at Hobby Airport (orange) and George Bush Intercontinental Airport (green). The two are correlated with an $R^2$ of 0.81. (Source: Data from NOAA)

monsoons, levels out and creeps back eastward. With the advection of winds aloft from the displaced low-pressure zone, the normal oscillations of the Pacific jet stream funnel into an enhanced storm track that channels low-pressure cyclones straight through the southern part of the United States, from southern California through the Colorado River Basin. These conditions also cause the Gulf to swell; the sea level rises and low-pressure systems emerge, leading to wetter conditions in the Gulf region, including east Texas (Kennedy et al. 2007).

As the City of Houston grew amid the humidity and storms, developers and city officials undertook to drain the swamp into a series of bayous, which course through the urban area to this day (Sipes and Zeve 2012). However, tropical storms bringing devastating storm surges pushed the city thirty miles inland, with the coastal City of Galveston bearing much of the burden and risk of commerce.

The primacy of Galveston in the region changed, however, with the massive hurricane of 1900. Destroying most of the city, the hurricane showed Texans that developing a major port city on the state's Gulf edge might prove folly in the long term. And the boosters who lived in Houston were eager to promote the trend, digging a lengthy ship channel from the coast into the Port of Houston, through which ships could carry large cargo (Brennan 2002). By bypassing the debilitated Galveston, the City of Houston got its first taste of economic success—a feeling that would only grow with the success of the local Spindletop oil field. As the oil boom ensued, driven by the growth of the automobile industry, Houston grew apace.

City planners saw in Houston a veritable laboratory for the modern automotive city. Rather than being compact, dense, and downtown centered, Houston would sprawl out for miles. Everything could be accessible, provided the roads were big enough to accommodate personal vehicles. Rather than building upward to house a clustered populous, Houston would build outward to afford the middle class the American Dream—the life of a private citizen, stereotyped in terms of a house, a car, two kids, and a wife (Melosi and Pratt 2007).

Yet, numerous hard problems came along with this schema, as repeated in Los Angeles, Phoenix, Arizona, and other modern metropolitan areas. Houston's growth consumed vast acreage of coastal wetlands, which provided crucial drainage for heavy rains, and valuable breaks for cyclones careening into the area from the Gulf of Mexico. The well-paved streets that proliferated provided impervious surfaces, optimal for streams of runoff to form and gush throughout the urban area. Furthermore, the bayous dug out to drain the marshlands would be reinforced with concrete, providing channels that would quickly overflow their banks during intensive floods, contributing to extensive flooding (Juan et al. 2020).

As the city continued to expand into the grasslands and fields of Southeast Texas, fed by the profits of the oil industry, more refineries sprang up from Baytown to Pasadena, ultimately claiming much of the entire southeastern area of the Houston area for the heavy petrochemical industry. Houston's growth deepened with its commitment to the oil industry, which created obvious hazards pertaining to natural disasters. Not only would urban development increase the flashiness of flooding in the area but also the increasing industrial sector posed the threat of toxic chemical contamination in the event of severe weather events such as hurricanes (Friedrich 2017).

It is partly for this reason that the growth of Houston's residential areas followed redlining codes, through which minority populations were restricted to areas more subject to pollution of the air and contamination of the land. While Houston maintains among the most diverse inter-racial middle classes in the United States, poor communities tend to be disproportionately comprised non-White populations. Consequently, cancer rates in Houston stand at much higher levels for non-Whites in what many activists consider a crisis of environmental injustice. It is, of course, no coincidence that these areas with higher health risks also coincide with more flood-prone areas (Sansom et al. 2022).

Intense rainfall events are not the only climate change–related problem confronting Houston. Aside from its famous hurricanes, Houston also catches its share of droughts. The worst drought on record came in 2010–2011, breaking records for the total number of days over 100 degrees Fahrenheit, along with average monthly temperatures and lowest 12-month rainfall. In some areas of the Upper Gulf Coast, water demand exceeded supply, leading to overreliance on groundwater and a concomitant trend of subsidence (Jiang 2015). The drought persisted until 2015, and overpumping caused sinking ground levels in an area already close to sea level, bringing about greater flood hazards. Thus, one disaster intensified the next. Meanwhile, the droughts can lead to stress on local ecosystems, causing habitat loss and death (Schmidt and Garland 2012; Nielsen-Gammon 2012).

The year 2022 marked more than 40 days of sudden triple-digit temperatures, bringing extreme drought to more than half of Texas in what Mozny et al. (2012) called a "flash drought, a very rapid decline in soil moisture during a 3-week period." Although the worst impacts were felt in the panhandle area up north, Houstonians were hit with brutal temperatures, causing water demand to skyrocket. Such conditions can cause major problems for harvests and livestock, as the grass that cattle eat withers and dies, forcing farmers to buy feed on the open market. Droughts also pose a significant threat of wildfires in the area and can lead to rivers drying up completely. On top of this, the droughts can draw the moisture out of the clay soils around Houston, pushing conditions past the wilting point and baking the ground to the point of hardening and cracking the city's aging water infrastructure (Dougherty 2022) (Fig. 12.3).

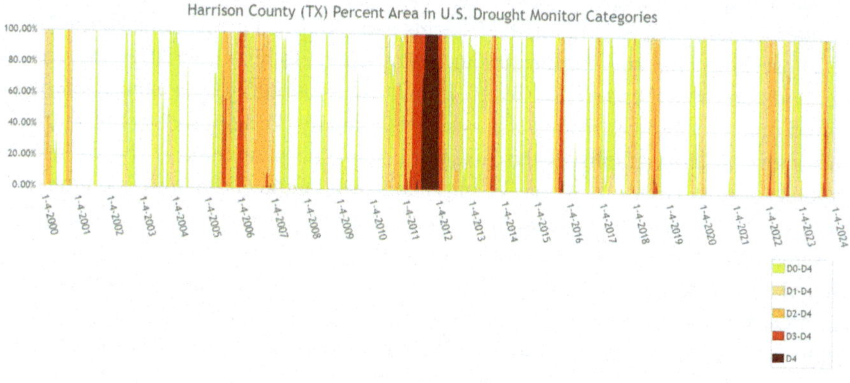


**Fig. 12.3** Historical drought in Houston. (Source: Drought monitor, https://droughtmonitor.unl.edu/DmData/TimeSeries.aspx)

So, as the oil industry continues to contribute to a changing climate in which severe weather events are only intensified, Houston's growth presents an existential conundrum. As all Houstonians are aware, their city's economic success—or, indeed, primacy—depends on the energy industry, which simultaneously contributes to hazards that bring about near-biblical devastation. While the palpable reality of that contradiction of interests presented therein can be felt at home, it remains true for the world at large. As the capitol of the US energy industry, and one of the top producers in the world, Houston can be viewed as one of the top global contributors to and beneficiaries of climate change, whose disproportionate impacts are felt in poorer countries throughout the world.

## 12.3 Ecological Impacts and Ecosystem Services

Despite its history of urbanization and natural disasters, the Houston area lies within the ecologically rich Texas coastline. While these biodiverse ecosystems face natural disaster threats accelerated by climate change, they can also play mitigating roles in the worst impacts of climate-related problems. For this reason, the restoration of crucial habitats around the Houston area has been cited as a key means of adapting to climate change.

Anyone who has been to Houston likely knows that the city prides itself on a multicultural melange of restaurants and culinary experiences—and many of those encounters with that melting pot of Vietnamese, Creole, Tex-Mex, Southern soul food, and other exquisite cuisines include oysters. Gulf oysters are unlike those found in the Pacific Northwest or the Atlantic seaboard; they are big, they are meatier, and Texans love them. And all of that is well because oyster reefs provide an excellent barrier to climate disasters (Hynes et al. 2022).

Oyster reefs offer what some environmentalists call a "living shoreline"—a thriving place for mollusks and other creatures that, in turn, provide happy hunting for birds and other animals that do their part to connect global habitats. These shorelines not only bring the world its life and animate nature, but they also act as buffers to massive waves that would otherwise destroy the coast. Thus, oyster reefs help to counteract the impacts of intensified storms and rising sea levels (Blackburn 2017).

Those birds that find sustenance in the oyster reefs also flourish in the coastal wetlands and marshes of East Texas, known internationally as a Mecca—albeit a humid one—for birders from as far away as Germany. Though it does not share the characteristic iconography of the Texas prairie of the panhandle or the vast desert expanse of East Texas, the seagrass and mangroves of the Gulf Coast could provide a key to unlocking climate adaptation (Guannel et al. 2016).

Of course, the whooping crane of the Gulf wetlands is a storied species in the United States, while other birds and aquatic animals (shrimp and so forth) thrive in the area, but the land–water interface also soaks in and stores carbon. Known as "blue carbon," this absorbed carbon could help to mitigate the rise of climate change through ecological means that support the ramping down of emissions, ultimately turning runaway feedback loops into manageable systems. Perforce, some studies show that mangrove blue carbon strategies scale best, offering the best national strategy to mitigate climate change (Taillardat et al. 2018). Like oyster reefs, the mangroves and wetlands of the Texas coast provide buffers to the hurricanes and tropical storms fed by the growing amount of water vapor caused by increased evaporation in a warming, subtropical climate.

Projects to restore Texas' bountiful oyster reefs, its coastal prairies and mangroves, and its other important ecosystems play an important role in local efforts to counteract the impacts of climate change in the Houston area; however, those efforts run against the consequences of climate change even in the process of working to defend against it. After Hurricane Harvey, the massive amount of precipitation dumped on the Gulf caused the salinity of coastal areas to decrease, leading to a crisis for oyster habitat (Du et al. 2021).

Meanwhile, sea level rise caused by climate change is shown to impact mangrove forests sensitive to inundation changes. Storminess can increase erosion and put a damper on plant productivity, as well as increasing sediment loads, which are also affected by shifts in precipitation regimes writ large. So, even as oyster reefs and coastal wetlands present an opportunity to save coastal cities from some of the worst climate impacts while contributing to the global campaign to diminish climate change before it gets out of control, the window to act is closing. The worse the climate change becomes, the less ecological restoration efforts will be sustainable amid the rising challenges that it portends (Blackburn 2017).

To mitigate the cost of flood events, the Storm Prediction, Education and Evacuation from Natural Disasters Center at Rice University called for a "nature-based economy" in the upper Gulf Coast by creating an environmentally protected area that could provide nonstructural relief from storms. The conceived Lone Star Coastal National Recreation Area, comprising 220,000 million acres, would provide revenue from tourism while also ensuring "blue carbon" sinks and flood breaks for decades to come. Inclusive of lands in four counties, the protected areas would include coastal estuaries, marshes, and prairies, along with barrier beaches, bayous, freshwater wetlands, and riverine hardwood bottomlands (Blackburn 2017).

Protection would offer agricultural and grazing lands along with recreational opportunities, parkland, and ecosystem services such as resilience and storage of storm surges, provisions for biodiversity promotion through hatcheries and feeding/nesting habitat, and offsetting the impacts of economic development while also improving water quality. Since some of these lands have fallen into degraded status, due to invasive species including wild pigs and tallow trees, or to disconnection from other habitats, great opportunities for enhancement also exist (Blackburn 2017).

There's a problem, though. Some 70–80% of this area is owned privately, and those landowners need financial incentives to transition toward conservation practices. Thus, advocates formed a market-based ecological exchange called the Texas Coastal Exchange to put an appropriate price tag on the lands that would otherwise be used for farming and other revenue-producing activities in an area increasingly broken up by urban development. Since a number of commercial and industrial stakeholders might benefit from such services, buyers in those sectors exist along with the gift market associated with bird watchers and other conservationists, as well as compliance buyers (Blackburn 2018).

Amid the practice of formulating appropriate financial incentives and ensuring the conservation and improvement of low-lying areas, the Texas Coastal Exchange grew into a body that could apply similar approaches to other areas around the state. Restoring wildlife and protecting the climate at the same time produced a popular groundswell in Texas, sparking a point of pride in local biodiversity and stimulating economic growth while also offering unique opportunities for local businesses to promote corporate responsibility. At the same time, the urban impacts of climate change require more urban responses—not only ecological conservation (Blackburn et al. 2015).

## 12.4 Urban Development

Wetland restoration is an easier sell in parts of the Deep South where market-based solutions hold sway over the intervention of "Big Government." At the same time, major hurricane events have brought even Conservative areas to agree to urban climate plans such as the 2020 Houston Climate Action Plan, or Resilient Houston. According to this plan, the four major arenas of action correspond to transportation, energy transition, building optimization, and materials management (City of Houston and Shell 2020).

Some of these strategies look like the same strategies for restoring gulf wetlands. Houston's bayous were engineered for the purpose of long-term canalization, not in the interests of the preservation of biodiversity or even the practical purposes of mitigating flood extent. However, by restoring wetlands along bayous and connecting them to floodplains, Houston can do both—adding habitat for migratory birds and other charismatic species while also decreasing the specter of mass destruction. Along with such restoration efforts, new green stormwater infrastructure on the docket includes bioswales, rain gardens, and green roofs (Scungio 2020).

Aside from highlighting the need for the restoration of habitat where possible, in order to capture and store carbon, the action plan calls for a transition to cleaner and more efficient energy sources that increase resilience. Houstonians were scarred by the aftermath of Hurricane Harvey, when flood waters became contaminated with toxic chemicals from refineries, creating a doubly dangerous situation with long-lasting impacts on land and water quality. In addition, worsening winter storms have brought about the need to upgrade and winterize power plants, while investing in the transformation of contaminated areas—called "brownfields" in developer parlance—into energy resources by installing renewable energy on sites unsuitable for other developments (Scungio 2020). For these energy purposes, the City of Houston announced a program called Resilient Houston in 2020, offering an update 2 years later (City of Houston and Shell 2020; City of Houston 2022).

For transportation, Houston hopes to transform the foundations along which the city's sprawling shape took over the landscape. Created in the early 2000s as part of a forward-thinking project to diminish fossil fuel consumption, Houston's light rail remained very limited in extent during the early 2000s. However, the Climate Action Plan promises to extend equitable and safe transportation choices, including light rail (Scungio 2020). Also, the City of Houston seeks to convert its bus fleet to an electric and low-emission fleet, decreasing the amount of pollution from its extensive public transportation network. Lastly, the city wants to emphasize decreasing mileage traveled for future urban developments in Houston, seeking to increase localization and availability (City of Houston and Shell 2020).

At the same time, Houston traffic remains among the worst in the nation, exhibiting the conflicting interests between short-term and long-term goals. Yet, reducing the number of cars on the street continually takes second priority to expanding the freeways in a process that would make Sisyphus sympathetic. One controversial highway expansion project in Downtown Houston will cost billions of dollars and tear down an apartment complex of high-end lofts near the baseball stadium, thus pitting the interests of motorists against those of downtown's denizens and stadium-goers, as well as climate advocates who want to decrease traffic rather than increase capacity.

To relieve traffic, developments need to emphasize walkability and increase available options on the local scale, but at the same time, new buildings can offer climate relief by increasing efficiency and adding renewable energy that can feed back into the power grid. But older buildings can also contribute to this potential through upgrades assisted by the city's action plan. Throughout Houston, urban gardens have spread in recent years to highlight opportunities for communities to come together locally and produce food while expanding green acreage within the city limits. While they seem small in the grand scheme of things, such green space attractions, combined with restoration efforts, on a city-wide level, contribute to carbon capture and mitigation of climate catastrophe.

Houston's plan for mitigating climate change also involves waste management. By reducing the amount of space taken up by open landfills, the city can reduce the methane emissions created by biodegradation. Also, the reduction of waste is proposed through city-sponsored composting programs, along with support for businesses that can cut waste through entrepreneurship. Lastly, a major problem in Houston is the presence of waste upstream, which moves downstream during major flood events. Cleaning up these sites helps ensure the protection of Houston's ecological sustainability.

At the same time, the nearby City of League City is tackling the problem of drought by building four lift stations to manage the processing and storage of water. This decentralization helps modulate the distribution of water in order to ensure that each area can continue water services in the event that part of the infrastructure breaks. In addition, new limits on groundwater reliance to 10% of the total water use have been set into place in order to prevent further subsidence. Meanwhile, treated wastewater is increasingly used to replace fresh water supplies used in irrigating golf courses and similar areas (Scungio 2020).

## 12.5 Class and Ecology

Although Houston's Climate Action Plan advances terrific steps to confront climate change using the means at the city's disposal, the layout of the city still punishes poorer, non-White areas with greater impacts from flooding. According to a recent study looking at federally overlooked flood inequalities in the city (Flores et al 2023), non-Whites are disproportionately exposed to high flood risk areas. Although the Action Plan makes important strides, it does not tackle issues of income inequality and the disproportionate burden of climate vulnerability. While Houston is, indeed, a multiracial and ethnically diverse city leading the United States in progressive efforts on a number of issues, its city plan still lacks consideration for real problems that are difficult to face in modern times, because they show that, despite the advances made, there is still much work to be done.

To assess the problem of poverty and flood vulnerability, we used a geographically weighted regression, which determines whether the clustering of low-income areas is a result of their location near 100-year floodplain, which are more likely to bear the brunt of intense storms (Fig. 12.4).

What we found was a fairly clear pattern: city blocks clustered around floodplains tend to be lower income and higher density. This clustering has a compounding impact, because flood insurance costs more in areas of repeated flooding,

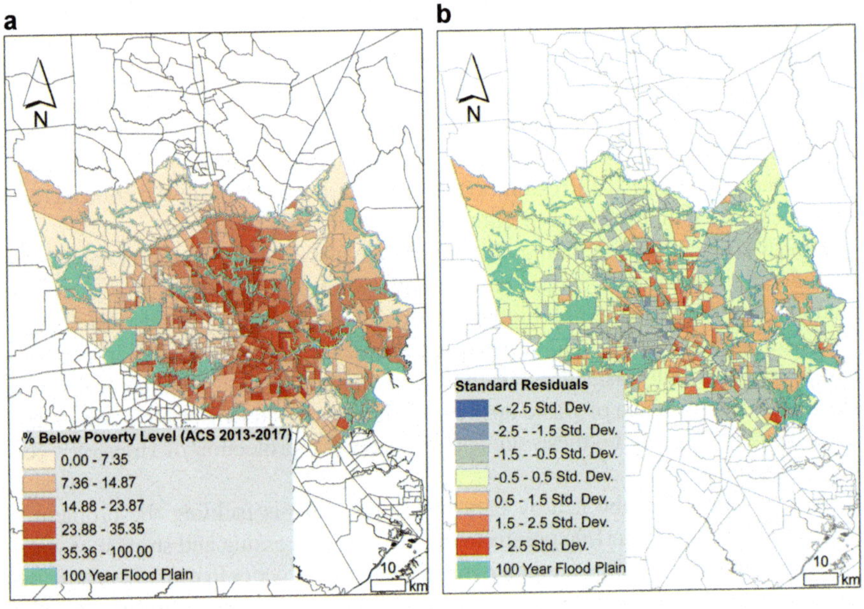


**Fig. 12.4** (**a**) Choropleth map showing the percentage of households under poverty level by census tract, (**b**) geographically weighted regression showing standard residuals of spatial correlation between 100-year floodplain, population density, and the percentage of impoverished households ($R^2$=0.33). Made using ArcPro with an Esri shapefile for the US Census tract demographics (2019) and City of Houston shapefile for 100-year floods (2023)

making it more onerous. Since fewer people in flood zones will have flood insurance, the probability of losing property due to inundation and its effects, such as mold, warping, and water damage, increases appreciably. As a result, people in lower-income areas face a structural disincentive to home and neighborhood improvement, trapping communities in depressed conditions where investment loses purpose.

## 12.6 Conclusions

The climate change challenges faced by Houston, Texas, involve a pattern of increasing temperatures and intensifying storm events, leading to floods. These growing patterns of floods and droughts create economic as well as social and political problems not just for residents of the city but also for wildlife and ecosystems in the surrounding areas. While the City of Houston, conservationists, and other stakeholders are working to develop creative and adaptive responses to these problems, Houston also remains at the center of the global energy industry, giving it the potential to become a leader in a transition away from those fossil fuel emissions that contribute the most to climate change.

Ecosystem services provided by habitat restoration and the conservation of wetlands and prairies near the coast offer a useful means to afford landowners, farmers, and other land users a substitute for economic development practices that might damage critical ecosystems that help mitigate flooding. By paying reasonable prices to settle protected areas, the Texas Coastal Exchange has worked with buyers and conservationists to protect hundreds of acres, thus securing "blue carbon" and counteracting both the causes and effects of climate change in one effort. The means of ecological restoration in the Upper Gulf Coast, then, offer a model for climate-aware practices that can become economically productive while also increasing the quality of life by adding pride to residents' sense of place.

Meanwhile, Houston's climate plan remains relatively young but ambitious. Inclusive of integrated water resources management, the plan plots out a path for future adaptive measures to include the reduction of traffic, water use, and harmful waste, while advancing carbon-conscious design in terms of both new developments and city planning. This effort also involves greater localism, further expanding the sense of place in this expansive, modern city, to emerge from neighborhoods where availability, walkability, and bike-friendly communities can be prioritized over the traditional primacy of freeways and big, gas-guzzling trucks.

Yet, Houston remains plagued by contradictions. Freeway expansions continue apace, knocking down newer buildings with extravagant funding that could otherwise go toward safer and sorely needed transportation projects. The Houston Light Rail has expanded over the past two decades but still lacks a western and a northeastern arm that could alleviate traffic from the Interstate-10—the site of the most congested section of highway in all of Texas. At the same time, there is still hope. Researchers have found that replacing just 35% of all Houston's gas and diesel cars with electric would cut the entire emissions budget of the city in half. Between reducing the number of cars with public transportation and increasing the fleet of electric vehicles, along with smart city planning and ecological restoration, Houston can turn from being in the top 10% of city carbon footprints in the world to balancing out its climate budget.

## References

Armenta T, Lane KE (2010) Tennessee to Texas: tracing the evolution controversy in public education. The Clearing House 83(3):76–79
Blackburn J (2017) A Texan plan for the Texas coast. Texas A&M Press, College Station
Blackburn J (2018) Houston at the crossroads: resilience and sustainability in the 21st century. Rice University's Baker Institute for Public Policy. https://repository.rice.edu/server/api/core/bitstreams/43620865-06ec-4c46-bb38-1ecbb6a0acb5/content. Accessed 5 Dec 2022
Blackburn J, Buono RM, Dunbar L (2015) Emerging legal issues in hurricane damage risk abatement. James A. Baker III Institute for Public Policy of Rice University, Houston
Brennan K (2002) The Galveston hurricane. Chelsea House Publishing/James A. Baker III Institute for Public Policy of Rice University, New York/Houston
Chretien JP, Anyamba A, Small J, Britch S, Sanchez JL, Halbach AC, Tucker C, Linthicum KJ (2015) Global climate anomalies and potential infectious disease risks: 2014-2015. PLoS Currents. https://doi.org/10.1371/currents.outbreaks.95fbc4a8fb4695e049baabfc2fc8289f
City of Houston (2022) Houston climate action plan. Houston, Texas
City of Houston (2023) COH 100 YR FLOOD PLAIN HARRIS FEMA. December 3, 2023. https://cohgismycity.opendata.arcgis.com/datasets/MyCity::coh-100-yr-flood-plain-harris-fema/about
City of Houston and Shell (2020) Resilient Houston
Dougherty M (2022) Houston drought conditions busting pipes throughout the city KHOU11, August 9, 2022
Du J, Park K, Jensen C, Dellapenna TM, Zhang WG, Shi Y (2021) Massive oyster kill in Galveston Bay caused by prolonged low-salinity exposure after Hurricane Harvey. Sci Total Environ 774:145132
ESRI, US Census (2019) 2019 census tracts. Learn_ArcGIS, Shapefile, owner
Flores AB, Collins TW, Grineski SE, Amodeo M, Porter JR, Sampson CC, Wing O (2023) Federally overlooked flood risk inequities in Houston Texas: novel insights based on Dasymetric mapping and state-of-the-art flood Modeling. Ann Am Assoc Geogr 113(1):240–260. https://doi.org/10.1080/24694452.2022.2085656
Friedrich MJ (2017) Determining health effects of hazardous materials released during Hurricane Harvey. JAMA 318(23):2283–2285
Guannel G, Arkema K, Ruggiero P, Verutes G (2016) The power of three: coral reefs, seagrasses and mangroves protect coastal regions and increase their resilience. PLoS One 11(7):e0158094
Hynes S, Burger R, Tudella J, Norton D, Chen W (2022) Estimating the costs and benefits of protecting a coastal amenity from climate change-related hazards: nature based solutions via oyster reef restoration versus grey infrastructure. Ecol Econ 194:107349
Jiang J (2015) Rapid land subsidence associated with severe drought: a case study in the Houston area, Texas. Doctoral dissertation, University of Huston, Huston, Texas
Juan A, Gori A, Sebastian A (2020) Comparing floodplain evolution in channelized and unchannelized urban watersheds in Houston, Texas. J Flood Risk Manage 13(2):e12604
Kennedy AJ, Griffin ML, Morey SL, Smith SR, O'Brien JJ (2007) Effects of El Niño–Southern Oscillation on sea level anomalies along the Gulf of Mexico coast. J Geophys Res Oceans 112:C05047

Melosi MV, Pratt JA (eds) (2007) Energy metropolis: an environmental history of Houston and the Gulf Coast. University of Pittsburgh Press, Pittsburgh

Mozny M, Trnka M, Zalud Z, Hlavinka P, Nekovar J, Potop C, Virag M (2012) Use of a soil moisture network for drought monitoring in The Czech Republic. Theor Appl Climatol 107(1–2):99–111. https://doi.org/10.1007/s00704-011-0460-6

Nielsen-Gammon JW (2012) The 2011 Texas drought. Texas Water J 3(1):59–95

Nielsen-Gammon J, Escobedo J, Ott C, Dedrick J, Van Fleet A (2020) Assessment of historic and future trends of extreme weather in Texas, 1900–2036. Texas A&M, Office of the Texas State Climatologist, College Station, Texas

Sansom GT, Fawkes LS, Thompson CM, Losa LM, McDonald TJ, Chiu WA (2022) Cancer risk associated with soil distribution of polycyclic aromatic hydrocarbons within three environmental justice neighborhoods in Houston, Texas. Environ Geochem Health 45(2):333–342

Schmidt DH, Garland KA (2012) Bone dry in Texas: resilience to drought on the upper Texas Gulf Coast. J Plan Lit 27(4):434–445

Scungio LA (2020) Houston climate action plan-quality of life committee-August 26, 2020. Huston, Texas

Sipes JL, Zeve MK (2012) The bayous of Houston. Arcadia Publishing, Charleston

Speer JB Jr (1980) Pestilence and progress, health reform in Galveston and Houston during the nineteenth century. Houston Rev Hist Cult Gulf Coast 2:120–132

Taillardat P, Friess DA, Lupascu M (2018) Mangrove blue carbon strategies for climate change mitigation are most effective at the national scale. Biol Lett 14(10):20180251

Trenberth KE, Cheng L, Jacobs P, Zhang Y, Fasullo J (2018) Hurricane Harvey links to ocean heat content and climate change adaptation. Abstract Plain Language Summary Key Points Earth's Future 6(5):730–744. https://doi.org/10.1029/2018EF000825

# Chapter 13
# Portland, United States of America

**Abstract** Considered one of the greenest cities in the world, Portlanders have enjoyed clean air and water with lush vegetation in and around town for decades. Portland is prone to winter and spring floods because the city is located at the confluence of the Columbia and the Willamette Rivers and Lain on Missoula flood deposits. The city shows highly flood vulnerable areas clustered in the historically developed areas in the lowlands that often serve low-income immigrants. Portland has improved stormwater management by a combination of grey and green infrastructure. With the intentional installation of storm green infrastructure in low-income and flood-vulnerable neighborhoods as well as a pioneering program such as a willing seller land acquisition program to move residents out of floodplains, the city strives to achieve its social and environmental equity goals. However, recent climate-related water hazards and development pressures challenge flood risk management. With a unique urban growth boundary containing growth within the limit, Portland has become denser. Such dense development, which often encourages development on floodplains, leads to uncoordinated efforts between land use and water planning, hindering a futuristic climate resilience plan that includes creating a flood-resilient city.

**Keywords** Smart growth · Green stormwater infrastructure · Floods · Willing seller land acquisition program · Portland

## 13.1 Introduction

In early September 2020 during the COVID-19 pandemic, Portlandians experienced the worst air quality in the world due to widespread thick smoke from nearby wildfires. The year before, the Portland Metro Area registered as the ninth best air quality in the United States, showing the debilitating impacts of these events. Combined with severe droughts and strong winds, the fires expanded rapidly throughout the Portland metropolitan region. Considered as one of the most destructive wildfires in Oregon, nearly 40,000 people evacuated with an additional half million people in



H. Chang, A. R. Ross, *Climate Change, Urbanization, and Water Resources*,
https://doi.org/10.1007/978-3-031-49631-8_13

the evacuation warning zone. During the fire incident, any outdoor activities were completely banned because of public health concerns. The wildfires resulted in killing 11 people, destroying more than 4000 homes and burning 1.1 million acres. Additionally, wildfires are believed to affect streamflow and water quality due to increased erosion rates resulting from less vegetation and hydrophobic soils.

In the following year, June 2021, Portland had a heat dome, with the temperature reaching a record-breaking 116 °F (46.7 °C). This extreme heat event was caused by an atmospheric record-breaking ridge of high pressure that brought anomalously warm air northward into the region. Since cloud development was inhibited by sinking air, solar heating was maximized during the summer solstice season. The ongoing extreme drought in the western United States further accelerated to lead to the historical heat wave (Chang et al. 2021). With this heat dome lasting for about a week, at least 116 people died due to heat-related causes in Oregon (of which 72 occurred in Multnomah County (Multnomah County 2021). Such excessive heat resulted in unprecedented increases in stream temperatures. Stream temperatures in Johnson Creek, an urban stream that goes through the City of Portland, recorded 28 °C, much higher than the threshold temperature for the survival of cold water species. These two extreme weather events exemplify how Portland can better prepare for climate-induced hazards to ensure its system is resilient to external shocks.

As one of the first US cities that the Climate Action Plan was initiated and implemented, Portland has been proactive in reducing greenhouse gas emissions and planning strategies adapting to climate change. In particular, the city has attempted to address social equity issues to improve the economic and environmental conditions of vulnerable communities within the city. Mitigation and adaptation strategies for reducing water-related hazards have also addressed the need for addressing broad social and environmental justice issues. With projected changes in climate and continued growth in population, the city still faces multiple challenges to achieve urban sustainability.

## 13.2 Climate and Geography

Located between the Cascade Range mountains (east) and the lower Coast Range mountains (west), Portland is divided by the Willamette River that flows from south to north. The current landscape of Portland is the result of the Missoula flood that happened nearly 15,000 years ago. As the ice dam broke in Glacier Lake Missoula, a vast amount of water flew down the Columbia River, inundating the east side of the Willamette River and even overflowing to the west hill areas of the Tualatin basin. The floods created fertile soils in and around the Portland metropolitan area while creating a condition that water cannot drain into soils quickly.

Portland receives approximately 38 inches (approximately 965 mm) of annual precipitation, with annual temperatures ranging from 2 °C to 29 °C. Characterized by the Mediterranean climate, Portland has distinct wet and dry seasons. The wet winter season receives more than half of the annual rainfall, but relatively warm and moist air from the Pacific Ocean makes its temperatures moderate. Snowfall in most city areas is uncommon, although Portland receives snow storms every 3–5 years, with annual snowfall amounting to around 10 cm. Large-scale atmospheric circulation patterns such as ENSO and Pacific Decadal Oscillation also affect Portland's climate. During El Niño years, when sea surface temperatures in the central and eastern equatorial Pacific Ocean are warmer than average, storm tracks migrate to southern California, and Portland is likely to have warm and dry winters and hot summers.

Portland's relatively flat topography in the east and the confluence with the Columbia River provided native people (Multnomah people and Cascade Indians) a place for fishing, hunting, and berry gathering for the native people. As the native population dwindled as a result of European contacts, Portland was initially a small stopping place along the Willamette River between Oregon City and Fort Vancouver. Since navigation was not ideal for reaching Oregon City with lower water levels during the summer, Portland, located at the confluence of the Willamette and the Columbia Rivers, grew fast as its location allowed it to have access to deep-draft vessels. During World War II, the city became a prime location for the shipbuilding industry, while it was a destination for war veterans after the war.

Like most US cities, Portland's population grew shortly after World War II. However, since the 1960s, with a new wave of suburbanization, population growth within the city has slowed, while the suburban population grew (Fig. 13.1). To prevent sprawl development from encroaching on fertile farmlands in suburban regions, Oregon initiated a new bill to contain development within the designated

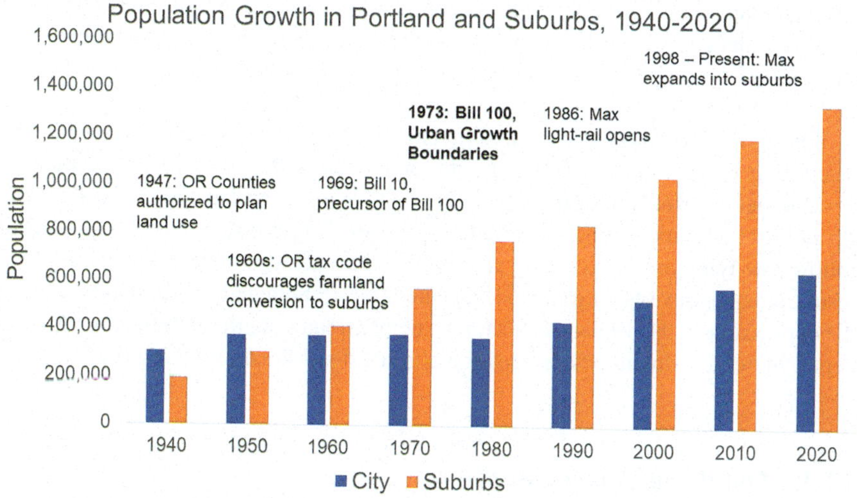


**Fig. 13.1** Population growth within the City of Portland and the Portland suburbs. (Original image available at http://energyfuse.org/wp-content/uploads/2018/02/portland1.gif was updated using the latest available information available; Source: 1940–2000 from U.S. census bureau and erik steiner, spatial history project, center for spatial and textual analysis, Stanford University (github.com/cestastanford/historical-us-city-populations) 2010 and 2020: US Census Bureau, decennial census (files SF1 and DHC, respectively))

**Fig. 13.2** An example of an urban stream with a confined width due to development along the stream, Johnson Creek, Portland, USA. (Source: Photo by author)

area. The unique land zoning systems, so-called urban growth boundaries (UGBs), have kept land development within UGB, even though the boundaries have expanded gradually since it was implemented. As a result, dense in-fill development has been encouraged to accommodate the growing population both within and suburban regions.

Portland's development was accompanied by the loss of streams and wetlands. According to Post et al. (2022), streams had disappeared well before residential development started since land conversion was already in place due to widespread agricultural activities. When interstate highways were constructed, they tended to follow low-lying areas (e.g., I-84 and I-205), encroaching on nearby streams. As a result, streams have been removed, rerouted, or straightened. Additionally, during the depression in the 1930s, such shortening of stream length, with fixed width and depth, increased stream power, creating chronic downstream flooding (Fig. 13.2).

## 13.3 Drinking Water Supply

As Portland's population grew substantially in the late nineteenth century, so did the need for new infrastructure development. The first major water project was constructing a drinking water supply system to deliver clean water to a growing population. Once relied on small urban steam for the main source of drinking water supply, the city needed to look for an alternative source outside of the city. The city started to source drinking water from the Bull Run Watershed, a protected area near Mount Hood, by creating a reservoir. Since its operation in 1895, the gravity-fed water now provides up to 212 million gallons per day. The Bull Run Reservoir now serves nearly one million people in the Portland metropolitan area. Considered one of the most pristine sources of freshwater drinking water in the United States, Portlandians enjoy crispy water. There have been several resolutions and ordinances to protect the drinking water source areas by prohibiting any commercial logging and other land management activities within Bull Run. The City of Portland has committed to maintaining it as the single source of drinking water, with no possibility of transfer of its rights to any private entity (City of Portland 2023). However, approximately 5000 households in Portland do not have piped water, which urban political ecologist Katie Meehan calls "plumbing poverty" (Deitz and Meehan 2019). These households are typically headed by people of color and low incomes, calling for action at the nexus of "water provision, housing, and social inequality" (Meehan et al. 2020).

Since the drinking water from the Bull Run Watershed is naturally corrosive, it has created a long-term problem for Portlanders with lead in their household plumbing. Residents with old pipes are particularly at risk of having the toxic metal leach from the pipes into their tap water. To alleviate this problem, the Portland Water Bureau built the Improved Corrosion Control Treatment facility to make the water supply less corrosive and protect these Portlanders from exposure to lead in their pipes. Approximately 15,000 houses built or plumbed between 1970 and 1985 are the most at-risk homes because this was an era when copper pipes with lead solder were common. Another concern is to comply with the state and federal water quality standards. After a few instances of bacterial contamination, the city has decided to install a new filtration facility that enables them to remove the microorganism Cryptosporidium. The project, which was contentious before approval due to its enormous costs and potential unnecessary need, is scheduled to be completed by 2027 (OPB 2022).

## 13.4 Flood Prevention Structure

Additionally, many water-related infrastructure projects were developed to reduce the risk of flooding. First, together with upstream dams that regulate flow, levees were constructed along the Columbia and Willamette Rivers to reduce flood risk. Built more than 100 years ago, the Portland Metro Levee System has a 27-mile stretch of levees managed by four drainage districts. Given that the four districts have independently maintained the levees in different sections of the river, the maintenance of levees has not been entirely efficient. Thus, these districts will be consolidated into one Urban Flood Safety and Water Quality District that improves efficiencies and better manage large-scale urban infrastructure. These levees protect major transportation and water infrastructure such as Portland International Airport and the Columbia South Shore Well Field, which provides drinking water for

Oregonians. In response to concerns about high water during flood events, raising the height of levees has been proposed in preparation for extreme flood events that are likely to be more common in the future with climate change.

Additionally, small dams and weirs were put in Johnson Creek, while the lower section of Johson Creek was rock-lined to control flooded water spilling over to the creek banks. While such flood control structures reduced flood risk in the upper basin, flood risks were transferred further downstream, inundating the lower section of the river. In particular, the East Lents area, which is located on a 100-year floodplain with lower elevation than riverbanks, has had a long history of flooding. Residents living in these flood-prone areas, typically low-income working class people, had a higher vulnerability to flooding compared to residents living outside of the floodplain (Fahy et al. 2019). Such a disproportionate exposure to flooding by different racial or income groups calls for environmental justice.

## 13.5 Droughts

While water shortage is not a severe problem in normal climate years, droughts do occur in Portland. Droughts can occur with different combinations of winter and summer precipitation and temperatures. For example, the 2001 drought was caused by unusually low winter precipitation, while the 2003 drought was caused by a combination of low summer precipitation and higher winter and summer temperatures (Bumbaco and Mote 2010). Together with a mild winter, the below-average winter precipitation in 2001 led to unprecedented low snowpack and streamflow in many streams around the Portland metro area. The exceptionally warm and dry summer in 2003 led to hydrologic droughts in most lowland areas. Since 2000, droughts have become more frequent; while there were only two major drought years in the 2000s, half of the years in the 2010s experienced severe (D2) to extreme (D3) droughts (Fig. 13.3). This coincides with rising temperatures and more extreme heat

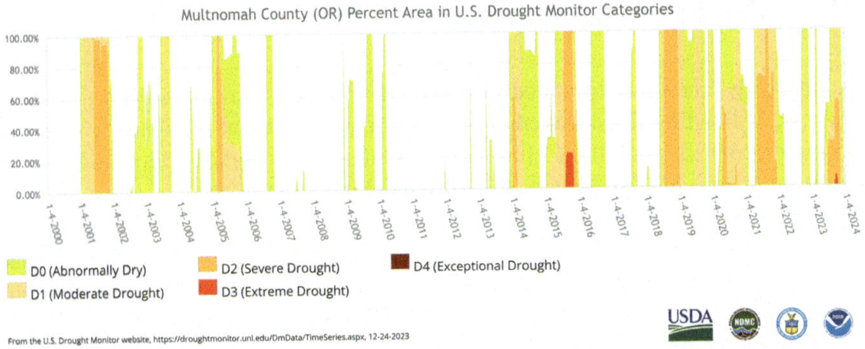


**Fig. 13.3** Drought severity in Multhoman County, 2001-2023. (Source: U.S. Drought Monitor: https://www.drought.gov/states/oregon/county/Multnomah)

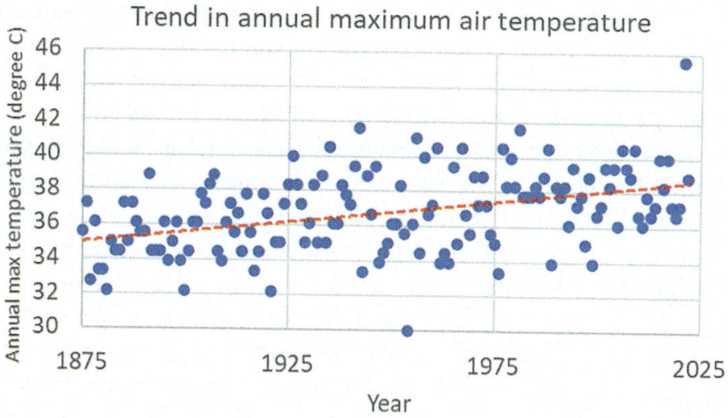


**Fig. 13.4** Portland annual maximum temperatures, 1875–2022. Each dot represents each year, and the dashed red line shows a long-term trend

events that have become more common in recent years. There is a statistically significant positive trend in annual maximum temperature in Portland (Fig. 13.4). The number of hot days (temperatures exceeding 90 °F) has increased since the 1970s. While fewer than 100 days had higher than 90 °F before the 1970s, the number of hot days was well above 100 days since the 1970s, with the highest number in the 2010s. More importantly, the 2010s saw a dramatic increase in the number of days that exceeded 95 °F.

When water levels from Bull Run drop during summer, Portland water is augmented by wells in the Columbia Slough. During the 1992 drought, when nearly all Oregon Counties declared a drought emergency, the city began restricting lawn watering. However, the drought did not last long.

With increases in summer temperature and decreases in summer precipitation, flash droughts have become more common in the 2010s (Fig. 13.4). According to global climate model projections, Portland's climate is further projected to have hotter summers and wetter winters. While precipitation projects are uncertain, these changes in precipitation distribution and rising temperatures are likely to result in more droughts, particularly summer-genesis droughts (Bumbaco and Mote 2010). Additionally, as extreme temperature events are more likely to occur more often in the future, like the heat dome that happened in June 2021, providing an adequate summer water supply would be further challenged.

## 13.6 History of Flooding

Portland has a long history of flooding. The notable flooding occurred in May and June of 1894, which resulted from a combination of unusually high rainfall amount and spring snowmelt from upstream areas. The Willamette River reached an elevation of 33.5 feet, inundating Portland downtown and surrounding agricultural lands for nearly 3 weeks. The floods brought huge damages, disrupting major transportation facilities, including the Union Pacific Railroad and two drawbridges that connected the east and west sides of Portland. The high water inundated 250 square blocks, destroying public, commercial, and residential buildings (Fig. 13.5). People fished in downtown streets while shopkeepers raised shelves above the floodwaters to reopen their businesses. Citizens were warned not to contact the contaminated water potentially containing bacteria from sewers.

The 1894 flood, regarded as the highest observed level of flooding on record in Portland, resulted in determining levee heights and other flood prevention measures by the US Army Corps of Engineers and other local agencies in the early twentieth century. While the construction of upstream dams and levees has reduced flood frequency, river discharge exceeded a few times in the twentieth century (Levee Ready 2022; Willingham 1983).

A notable flood was the 1948 flood when the City of Vanport, built on a floodplain and the second largest city in the region at the time, vanished as a result of a Dike break along the Columbia River. The city was developed to support a shipyard industry during World War II. With labors drawn from all over the United States, the city had a disproportionately higher number of Black people (MacColl 1979).

The flood was caused by rapid snowmelt with warm temperatures and heavy precipitation in late May and early June. The unusually high snowfall in the interior Columbia River during the winter of 1947–1948, in combination with relatively

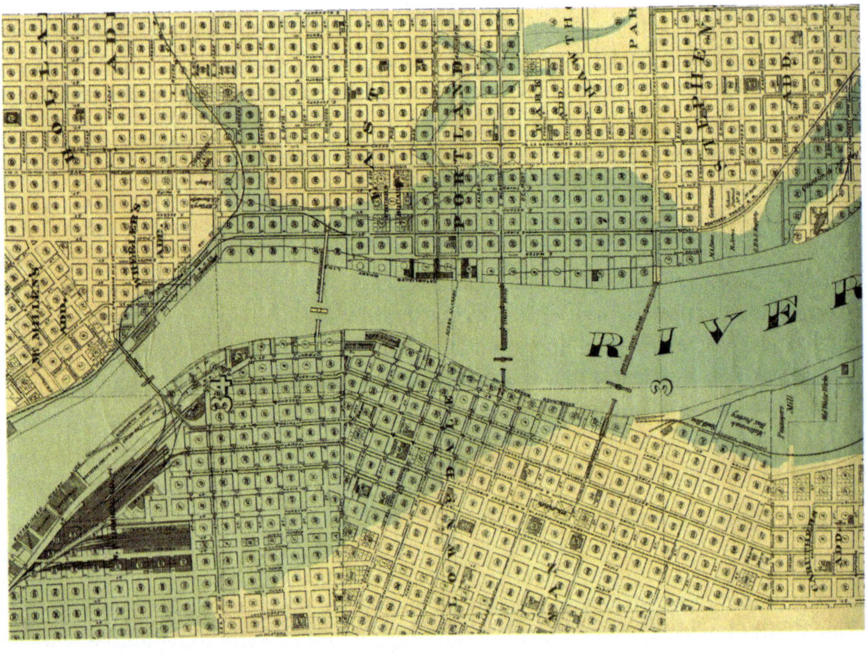


**Fig. 13.5** The spatial extent of flooding in downtown Portland during the June 1894 flooding. (Image credit: Levee Ready 2022)

cold spring temperatures, created an ideal condition for deep snow that lasted until mid-May. With a sudden shift in weather patterns in mid-May, warm moist air, originating from the Pacific Northwest, brought rain and thunderstorms. As a result, snow melted suddenly, raising the Columbia River's water level rapidly. By May 25, Portland's flood stage reached 7 m, 2.4 m above the flood stage. On the early morning of May 27, levees protecting Vanport failed, completely destroying the city. With the city remaining underwater for several weeks, the flood made nearly half of Vanport residents homeless (McGregor 2003).

Following the 1948 floods, the new Flood Control Act of 1950 was passed, allowing the construction of several dams and levees along the Columbia River, which has lasted to date. Additionally, a joint commission comprising US and Canadian representatives was created to address flooding issues in the Columbia River, resulting in the Columbia River Treaty. While the treaty has been somewhat successful in reducing flood risks downstream of the Columbia River basin, it did inundate other upland areas, including Canadian provinces. Given that some of the areas belong to native people in both the United States and Canada, the Treaty is up for negotiation.

The third biggest flood on record is the so-called Christmas Flood of 1964. The flood was caused by unusually cold weather with 1 m of snow in mid-December and sudden increases in temperature from atmospheric river. Since soils were frozen, melted snow runoff rapidly, raising river levels. The Willamette River in Portland reached 29.80 feet, recording the third highest level on record after 1894 and 1948. The 1964 flood had enormous impacts socioeconomically and geomorphologically. More than 7000 homes were damaged in Oregon, while channel paths shifted with scoring valley bottoms and depositing sediments downstream. Several bridges in the Willamette River were hit by log rafts, severely damaging one of the bridges and closing it for a year. According to the US Army Corps of Engineers, the damage could have been much worse had there been no seven dams that regulated flow (USACE 2022; Wikipedia 2022).

Another big flood occurred in Portland in February 1996. Similar to the 1948 flood, the flood was caused by a combination of heavy rainfall in previous months and heavy snow in January (some mountainous regions over 200% above average), followed by warm temperatures in February, which resulted in rapid snowmelt in the region. As the Willamette River water level reached 28.55 feet and Portland's seawall was not able to hold water, many city and citizen volunteers constructed a temporary wall on top of the seawall to prevent flooding in downtown Portland. While dams along the Willamette River effectively reduced flood risks by lowering the river water level by 2 feet, more intense urban development along the river and logging activities in the upper parts of the basin contributed to higher damages.

With anticipated climate change and sea level rise, Portland is not immune to flooding. Wet season precipitation intensity, particularly at the beginning (October) and the end (March) of the wet season, has already increased based on observed hourly data between 1977 and 2016 (Cooley and Chang 2021). According to a recent study that examined compound flood hazards resulting from increases in upstream flow and sea level rise scenarios (Helaire et al. 2020), Portland's water level is projected to increase by 0.78 m with a 10% increase in river run-off from upstream regions, while it is projected to rise by 0.12 m with a 0.6 m sea level rise scenario. When sea level rise and runoff increase scenarios are combined together, the river water level is projected to increase further by 0.87 m, suggesting that river systems respond to changes in river discharge, sea level, tides, and storm surge in a nonlinear way.

## 13.7 Dense Development and Residential Water Demand

While climate change is projected to increase summer water demand in the future, one way to reduce residential water demand in urban areas is through the manipulation of land uses. In other words, the density of residential development matters in water consumption. Dense neighborhoods typically have fewer lawns than sprawled suburban neighborhoods, resulting in reduced outdoor water use, which is the major water use in the summer. Within the City of Portland, a strong spatial variation exists in residential water use. Together with different sociodemographic characteristics of census block groups (CBG), densely developed CBGs (North and central Portland) show much lower water uses compared to upscaled sparsely developed high-income neighborhoods (Northeast and Southwest). Additionally, these affluent and educated neighborhoods not only use more water per capita, they also show higher sensitivity to temperature, often two or three times higher than densely developed neighborhoods. However, recent conservation and climate adaptation efforts have encouraged the adoption of water-efficient indoor equipment and native plants that do not require watering during the hot summers (Chang et al. 2017).

In the Portland metropolitan area, future projections of water demand are substantially different in two different neighborhoods—one with a denser and older neighborhood and the other with a sprawl newer neighborhood. The newer suburban neighborhoods exhibit much higher water demand, particularly during the summer, while the older urban neighborhoods have much lower water demand. Under the highest warming Hadley climate change scenario, the low dense suburban neighborhood has nearly 2.5 times additional water demand in June, compared to the densely developed urban neighborhood (Parandvash and Chang 2016).

Additionally, residents' water use behaviors are closely related to outdoor water use in Portland. According to Straus et al. (2016), residents' pro-conservation attitudes are highly correlated to summer water savings even after controlling for property size and other socioeconomic factors. In particular, landscape practices and the adoption of conservation technology are significant factors affecting residential summer water uses in Portland. In reality, many Portland residents voluntarily convert residential lawn grasses to native plants that do not require watering during hot summer days (Fig. 13.6). Additionally, water providers offer rebate incentive programs when residents replace old toilets with water-saving toilets to promote water conservation when water levels decline during late summer.

**Fig. 13.6** An example of a residential yard converted from lawns to minimize the use of water. (Source: Photo by author)

## 13.8 Floodplain Restoration and GSI Installation

Portland has implemented some pioneering practices in reducing urban flooding. For example, the willing seller land acquisition program, initiated by the City of Portland, was to move residents and businesses out of four flood-prone areas of Johnson Creek in the City of Portland (City of Portland 2019). Primary funding for the program comes from the city's Bureau of Environmental Services' capital funds. Acquisition and restoration were also supported by partnerships with Portland Parks & Recreation, the Metro regional government's open space program, and federal grants. Compensation to sellers was based on certified appraisals of the properties. In the East Lents area, the city purchased 60 houses over more than a decade. Once the properties were purchased, the city converted the site into a natural area, called the Foster Floodplain Natural Area restoration project. During the design process, the city held several open forums with residents to ensure the restoration project was well designed with inputs from diverse stakeholders. As part of the restoration project, wetlands, floodplain terraces, and other open spaces were constructed to hold water and sediments during flood events. In contrast, streets have been removed (Fig. 13.7). A hydrodynamic simulation study shows that the restoration project effectively reduces peak flow and sediment delivery downstream (Ahilan et al. 2018). For example, when a storm hit the area in winter 2015, the creek reached its record peak. However, inundation of the adjacent areas of the creek was limited compared to historic flooding in the area. Residents using restored floodplains that now have trails also perceive the positive value of restoration in reducing flood risk (Hong and Chang 2020).

**Fig. 13.7** Comparison of fostered floodplain areas before (left) and after (right) restoration. (**a**) Before restoration (1998), (**b**) After restoration (2018). (Image source: Google map)

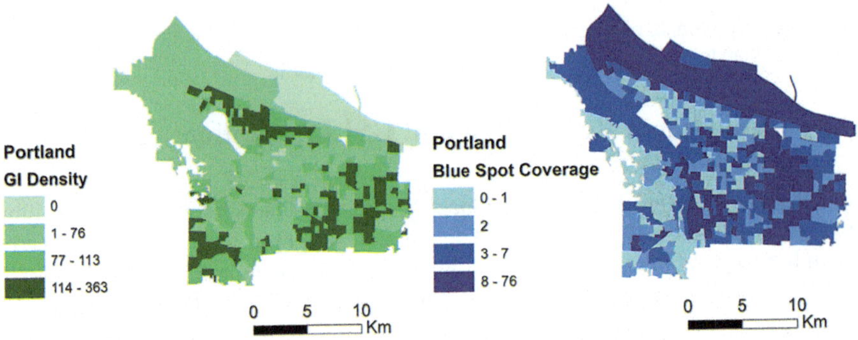


**Fig. 13.8** Location of potential areas of urban pluvial flooding as measured by % blue spot coverage and green infrastructure (GI) density at the census block group scale in Portland. (Source: Pallathadka et al. 2022)

The City of Portland has also been active in installing green stormwater infrastructure (GSI) to reduce storm runoff. Nature-based solutions in the form of GSI have become popular in Portland in the last two decades. When the city initiated big pipe projects to curb combined sewer overflow problems in the late 2000s, the city set aside a portion of its budget to install GSI. The GSI project, the so-called Tabor to River (T2R) program, focused on Southeast Portland, a relatively less affluent neighborhood (Fig. 13.8). This was an attempt to reduce the risk of urban pluvial flooding while addressing social equity by intentionally installing GSI in low-income neighborhoods (Baker et al. 2019; Pallathadka et al. 2022). The project included 3500 trees, 500 vegetated stormwater facilities, and over 100 private stormwater facilities, in addition to repairing or replacing over 260 km of sewer pipes.

Together with infrastructure changes, the program also developed a community outreach program that engages with the affected community via workshops and educational materials. According to the social survey, residents' awareness and perceptions of the GSI project vary by socioeconomic status, with higher income and educated residents participating more in stewardship actions (Shandas 2015). Such findings suggest that additional efforts are needed to reach out to people with fewer resources. In some instances, residents view GSI as inconvenient infrastructure, taking their parking space away and often piling up trash. Thus, the long-term effectiveness of GSI in reducing storm runoff and improving water quality could be questionable if attention has not been paid to adequate maintenance of GSI at the neighborhood scale.

## 13.9 Conclusions

As one of the world's most pioneering progressive green cities in Marine West Coast climates, Portland has enjoyed crisp and plentiful water for years. The unique 50-year history of urban growth boundaries also promoted dense infill development to prevent the spillover of land development to prime soils in the surrounding region. However, the city also faces water sustainability challenges associated with legacy development, growing population, and climate change. Most flood-vulnerable neighborhoods are located in and around disappeared streams and low-lying areas. The dense infill development has often not been coordinated with water planning, resulting in excessive water on streets. With projected increases in winter precipitation intensity, which usually coincides with the foliage season, the city needs to implement innovative stormwater management (Cooley and Chang 2017). A few promising projects have already been implemented, such as moving residents out of floodplains via a bond-sponsored willing seller land acquisition program, installing green storm infrastructure in low-income neighborhoods, and creating one entity to oversee flood protection structures. Combining social, ecological, and technological approaches in addressing flood risk can offer promising hope for flood resilience.

## References

Ahilan S, Guan M, Sleigh A, Wright N, Chang H (2018) The influence of floodplain restoration on sediment dynamics in an urban river. J Flood Risk Manage 11:S986–S1001

Baker A, Brenneman E, Chang H, McPhilips L, Matsler M (2019) Spatial analysis of landscape and sociodemographic factors associated with green stormwater infrastructure distribution in Baltimore, Maryland and Portland, Oregon. Sci Total Environ 664:461–474

Bumbaco K, Mote PW (2010) Three recent Flavors of drought in the Pacific northwest. J Appl Meteorol Climatol 49(9):2058–2068. https://doi.org/10.1175/2010JAMC2423.1

Chang H, Bonnette M, Stoker P, Crow-Miller B, Wentz E (2017) Determinants of single-family residential water use across scales in four western US cities. Sci Total Environ 596(597):451–464

Chang H, Loikith P, Messer L (2021) The June 2021 extreme heat event in Portland, OR, USA: Its impacts on ecosystems and human health and potential adaptation strategies. J Extreme Events 8(3):2175001

City of Portland (2019) Willing seller program. Accessed 11 Dec 2019. https://www.portlandoregon.gov/bes/article/106234

City of Portland (2023) About Portland's water system. City of Portland. https://www.portland.gov/water/about-portlands-water-system
Cooley A, Chang H (2017) Precipitation intensity trend detection using hourly and daily observations in Portland, Oregon. Climate 5(1):10
Cooley AK, Chang H (2021) Detecting change in precipitation indices using observed (1977–2016) and modeled future climate data in Portland Oregon USA. J Water Clim Change 12(4):1135–1153. https://doi.org/10.2166/wcc.2020.043
Deitz S, Meehan K (2019) Plumbing poverty: mapping hot spots of racial and geographic inequality in US household water insecurity. Annals of the American Association of Geographers 109(4):1092–1109
Fahy B, Brenneman E, Chang H, Shandas V (2019) Spatial analysis of urban floods and extreme heat potential in Portland, OR. Int J Disast Risk Reduct 39:101117
Helaire L, Talke S, Jay D, Chang H (2020) Present and future flood Hazard in the lower Columbia River estuary: changing flood hazards in the Portland-Vancouver Metropolitan Area. Geophys Res Lett Oceans 125(7):e2019JC015928
Hong C, Chang H (2020) Residents' perception of flood risk and urban stream restoration using multi-criteria decision analysis. River Res Appl 36(10):2078–2088
Levee Ready Columbia (2022) The great flood of 1894. https://leveereadycolumbia.org/timeline/22691/#:~:text=June%207%2C%201894%20%2D%20The%20Great,on%20record%20to%20this%20day. Accessed 23 June 2023
MacColl EK (1979) The growth of a city: power and politics in Portland, Oregon 1915–1950. The Georgian Press, Portland. ISBN 0-9603408-1-5. https://www.oregon.gov/biz/newsroom/pages/success_stories/portland-metro-levee-system-receives-upgrades-to-protect-city-from-future-flood-events.aspx
Meehan K, Jurjevich JR, Chun NM, Sherrill J (2020) Geographies of insecure water access and the housing–water nexus in. US cities Significance Proceedings of the National Academy of Sciences 117(46):28700–28707. https://doi.org/10.1073/pnas.2007361117
McGregor M (2003) The Vanport flood. https://www.oregonhistoryproject.org/articles/essays/the-vanport-flood/#.Y91JeXbMLuq
Multnomah County (2021) June 2021 Heat Wave. https://www.multco.us/help-when-its-hot/june-2021-heat-wave
Oregon Public Broadcasting (OPB) (2022) Portland facility to treat water for lead comes online. https://www.opb.org/article/2022/05/12/new-portland-facility-treats-water-for-lead/
Pallathadka A, Sauer J, Chang H, Grimm N (2022) Urban flood risk and green infrastructure: who is exposed to risk and who benefits from investment? A case study of three U.S. Cities. Landsc Urban Plan 223:104417
Parandvash GH, Chang H (2016) Analysis of long-term climate change on per capita water demand in urban versus suburban areas in the Portland metropolitan area USA. J Hydrol 538:574–586. https://doi.org/10.1016/j.jhydrol.2016.04.035
Post G, Chang H, Banis D (2022) The spatial relationship between patterns of disappeared streams and residential development in Portland, Oregon, USA. J Maps 18(2):210–218
Shandas V (2015) Neighborhood change and the role of environmental stewardship: a case study of green stormwater infrastructure in the City of Portland (OR, USA). Ecol Soc 20(3):16
Straus J, Chang H, Hong C-Y (2016) An exploratory path analysis of attitudes, behaviors and summer water consumption in the Portland metropolitan area. Sustain Cities Soc 23:68–77
The US Army Corp of Engineers (2022) The Christmas flood of 1964. https://www.nwp.usace.army.mil/Missions/Flood-Risk-Management/1964-Flood/#:~:text=The%20December%201964%20flood,extended%20through%20five%20northwestern%20states. Accessed 22 Dec s2022
Wikipedia (2022) Christmas flood of 1964. https://en.wikipedia.org/wiki/Christmas_flood_of_1964. Accessed 15 Dec 2022
Willingham WF (1983) Army engineers and the development of Oregon. Government Printing Office, Washington, DC. https://www.oregonencyclopedia.org/articles/willamette_flood_1894_/#.Y9GBGnbMLuo

# Chapter 14
# Conclusions

**Abstract** The 12 case cities show climate change and urbanization substantially impact water-related hazards. The mitigation strategies that each city has dealt with these hazards differ substantially, reflecting its complex nexus of social, ecological, and technological systems. Historically, uneven development and uncoordinated land use planning in developing countries have often resulted in city-wide high vulnerability to floods and droughts. In developed countries, vulnerable areas are also rooted in the historical development of the city, which is tightly coupled with economically disadvantaged and/or racially minoritized groups. Progress has been made in reducing water-related hazards through social learning by embracing ecological and social dimensions of vulnerability in recent years. However, there is still room for improvement in developing resilient, just, and equitable water-related hazard management strategies, given that climate-related water hazards may intensify in the coming decades.

**Keywords** Water-related hazards · Case studies · Lessons · Adaptation · Integration

Our studies show that climate change has broad and sweeping effects worldwide that manifest differently in different places. Yet, around the world, the increasing frequency of extreme events and ongoing urbanization make cities a crucial focus of both impacts and adaptation. While each case study stands on its own as a distinct model of particular interest in itself, together they show the utility of comparative analysis in parsing lessons on the impacts of climate variability and change, as well as urbanization, and how to address them. The variegated spatial and temporal array of changes, intensifications, and extremities require conceptualizing the problem as highly complex, demanding the flexible coordination of responses across multiple scales, from governance strategies to community-based efforts to technological development.

Floods and droughts generally create the most severe problems, requiring a greater depth of thought than simply increasing water storage. Droughts often create general impacts throughout large areas for a prolonged period, requiring large amounts of systemic actions that can include technical water transfers requiring


H. Chang, A. R. Ross, *Climate Change, Urbanization, and Water Resources*,
https://doi.org/10.1007/978-3-031-49631-8_14

existing infrastructure and time. In this sense, drought management requires forward-thinking policies focused on groundwater recharge, emergency water access, and aquifer protection. In addition, societal aspects of drought mitigation depend on individual responsibility and restraint, involving an ethical commitment to the community and a willingness to sacrifice amenities for its sake. All of these things align with social learning, the capacity of groups of people to develop together habits of resilience to climate impacts; repeatable practices of water conservation, and use that function to curtail the harmful effects of drought.

Like droughts, floods can impact large numbers of people, particularly in cities with dense concentrations of residents, but their effects are often felt as a pulse, or short-term, process rather than a press, or more continuous, long-term process. Floods shock and overpower, causing tremendous physical damage to the living environment and to the human population. Therefore, flood management depends on the implementation of existing infrastructure to remove as much water from the city as possible as soon as possible with great efficiency. This kind of infrastructure can include blue-green infrastructure, sustainable urban drainage system (SUDS), and other urban landscape features, but it also involves restoration of floodplains, catchments, and other ecosystemic approaches. For this, thinking about the city in an adaptive relationship with urban ecological systems helps to address multiple climatological problems at once, from carbon capturing by growing more trees to rainwater infiltration by increasing green spaces such as rain gardens and bioswales.

Thus, floods and droughts call for different forms of infrastructure with two different temporal levels in mind—the former as a pulse effect and the latter as a prolonged process effect. Both forms of climate hazards are set to increase in magnitude, meaning that their socioeconomic impacts will worsen in the future. Hence, the need for better planning and international coordination will ultimately prove more cost-effective than putting off preparations. Because the history of hydrological development suggests that development and expansion typically follow unmanageable crises, one of the greatest difficulties lies in convincing policymakers to act now, rather than wait for the next crisis to necessitate further development. On top of this, the impacts of catastrophic climate events virtually always fall on the most vulnerable, who are generally poorer, marginalized people. Uncoordinated planning and unequal development exacerbate the vulnerability of these people and feed back into broader social problems. The question then remains, "How can we create a just and equitable future to promote urban resilience to extreme events?"

The answer can only arise in a framework that enables adaptive networks to shape development in accordance with the need to mitigate the phenomena that cause climate change while simultaneously improving communities' ability not just to "bounce back" but to utilize today's crisis for tomorrow's solutions. This means connecting land use to water policy and climate action across all sectors and neighborhoods. This connectivity in space and time, between generations as well as levels of administration involving multi-scalar networks of governance, society, and ecology, enhances democratic engagement, decreasing vulnerability while also promoting social learning based on rational responses to material problems.

While climate change poses dire consequences for all of humanity, responding to it offers immense potential for collaboration that could ameliorate existing polarizations by linking people in common cause. Together, people can coproduce knowledge and create transdisciplinary approaches, involving academia together with other institutional vectors that operationalize participatory planning and design throughout the process. This codesign and coproduction of knowledge and frameworks emerging from an understanding of social-ecological-technological systems can enhance the analysis of connectivity and cascading impacts, feedback, and wicked problems typical of complexity. We know as scientists that some of the most successful governance models implement social-ecological-technological systems, such as smart cities and water-sensitive cities. We just need to implement those models in ways that suit the needs of distinct places around the world (Table 14.1).

Based on our findings, it seems clear that those cities of the developed world with more resources and freedom from deprivation tend to rely more on engaged citizenry, while the prevalence of economic plight tends to make citizen-led approaches more difficult. Yet, the availability of resources does not necessarily indicate the equity of city planning. Cities such as São Paulo and Istanbul enjoy far greater economic wealth than cities such as Newcastle or Portland in total, but per capita remains much lower. Thus, poverty tends to be present in greater extremes where the population outstrips even the tremendous productivity of these urban economies. The inequitable distribution of wealth in areas of high population parallels the top-heavy forms of management, as the masses of people become more difficult to organize. Yet, community-based efforts in Lagos and other megacities provide insight into ways that plans developed on smaller scales can be reproduced and coordinated more widely.

Implementing new standards for disaster policy amid climate change is not easy, but as we have shown, all cities strive in some way toward the same goal. Each case study illustrates the fact that climate and water policy is more than residents; it requires cooperation between public and private institutions for blue-green infrastructure and sustainable urban drainage systems, eventually providing a greater sense of investment and a better sense of settlement. Like Newcastle's Learning and Action Alliances, collaborative groups can be more informal and produce their own cultures with relationship building emerging from the satisfaction of unique needs. Collaborative organization creates new entities, puts life into neglected or hidden systems, to bring together ways to approach these goals.

The great thing is that, while efforts like blue-green infrastructure and SUDS have been tested and proved in action, they remain relatively new. There is so much promise that renewed efforts in different places hold. We have seen how an idea developed in Melbourne, Australia, spread to Seoul, Republic of Korea, taking on new aspects and creative adaptations. With the growth of smart cities, artificial intelligence, and other technological innovations, who knows what future efforts of climate adaptation could bring? As São Paulo shows, albeit in a nascent sense, sometimes the logic of keeping water out of the system can be altered toward making use of water features and reintegrating water into the landscape in ways that prove aesthetic as well as utilitarian. By installing rain gardens, water parks, and

**Table 14.1** A summary of flood and drought mitigation strategies in 12 case cities identified by the authors

| City | Water provider | Flood mitigation strategies | Drought mitigation strategies |
|---|---|---|---|
| Seoul | City of Seoul | Underground tank, stream restoration, social education, BGI | Water conservation |
| Jakarta | Jakarta provisional government + private operators ⇒ City (2019) | Barrier techno-island, slum clearances, moving capitol city | Crop changes |
| Istanbul | Istanbul Public Utility (ISKI) | New dams, reservoirs, potential for WSC planning | Expanding reservoirs, expanding water pipelines |
| Newcastle | Newcastle | BGI, runoff attenuation features (reduce flood by 30%) | Natural engineering |
| Barcelona | Sociedad General de Aguas de Barcelona | Water stays out of the city (similar to Portland), sustainable urban drainage system (SUDS) | Desalination |
| Lagos | Lagos State Government | Community engagement with slum populations about flood mitigation in low-lying areas | Expanding water sources |
| Cape Town | City of Cape Town water service | Infrastructure upgrade | Strict regulation of water supply, social learning |
| Melbourne | Melbourne Water (Government of Victoria) | "Water-sensitive cities" (WSCs) | Differential spatial response and farm capacity building |
| São Paulo | Federal and local | Artificial basins, linear parks | Stream restoration |
| Mexico City | Both federal and local agencies | "Water squares" | Surface water use, infrastructure |
| Houston | City of Houston | Hurricane threat, wetland rehabilitation, canalization (overflow canal) | Consumption reduction |
| Portland | Portland Water Bureau (City of Portland) | Combination of gray & green infrastructure to mitigate pluvial flooding, willing sellers program (relocation of people on flood-prone areas), blue-green infrastructure (BGI) | Demand management, water conversation, new sources, coordination among water providers |

and fountains, we can be creative with the sustainable use of abundance while ensuring security in times of great need.

Form follows function but does not have to be displeasing. The function of water policy often emerges in relation to governance and ownership, and if the objective is to serve the city and its people, the method of organization ought to reflect that. Private ownership over water resources can succeed in some ways, but it also tends to underserve more vulnerable parts of the community, thus reinforcing feedbacks of unsustainable informal habitations and other spatial manifestations of poverty cycles. The debate over municipalization in Barcelona illustrates potential transformations in water governance that could generalize the benefits of adaptive management by investing responsibility in larger portions of the populace without detracting from the need for multi-scalar networks to collaborate for robust approaches to the myriad hazards that climate change poses.

Precisely because of the multitude of hazards, adaptive measures must be specified locally, involving the integration of surface and groundwater management (especially groundwater recharge) with the help of AI-enhanced models for locating hotspots for flooding or urban warming. Hence, modern technology can make social learning more adaptive, rather than developing and reinforcing systems based on models that fail to fit reality. The social-ecological-technological systems (SETS) framework suggests that social assemblages tend to coevolve with technological and ecological systems, so the avoidance of maladaptive use of technology will also feed into better societies while improving ecological conditions. The nascence of multi-scalar connectivity in the different geographic settings of global cities offers a tremendous, generational opportunity to mold the best, most materially relevant networks possible, from how we think to how we act to what we do.

In a similar way, cities across different continents face similar challenges from socio-technical teleconnections, but these also provide opportunities to enhance predictive capabilities, using geospatial technology to inform and enable responses to weather and climate phenomena. The projection of meteorological events, the facilitation of emergency warning systems, the construction of international, reflexive infrastructure to provide support when and where it is needed—all these can and must be mobilized internationally in the spirit of cooperation.

Our studies found that functionality tends to follow engagement of localized civil society, vis-a-vis civil organizations, governmental institutions, and industry collaborating with academia to encourage active participation of all parties, including the population. This means involving representation of civic interests and placing them on the same level as those of other stakeholders, ensuring that potential energies are realized rather than squandered on projects that only serve the few. There is a sea change in the civil engineering tradition as it relates to hydrologic infrastructure, introducing the public to processes that were long placed out of their reach. However, more vulnerable populations tend to be less available for participatory projects, rendering clear the need to promote equity as part of adaptive efforts.

Historically, the development of infrastructure carries a long legacy effect, as demonstrated by our study of cities, whether they are flood-related infrastructure or restoration. New designs must be created in view of long-term impacts with explicit engagement of diverse stakeholders, as the results of short-term development often prove more expensive over time. Fragmentation has resulted from ad hoc processes of meeting immediate, relatively narrow needs fixing specific problems in place and time, and reconciling those staggered institutions can become highly useful in improving efficiency, clarifying overall problems and developing articulated solutions. Today, Portland offers a way forward, with one entity overseeing flood-related management throughout the city. Globally, the creation of unified efforts makes collaboration across cities and countries easier, as well.

Much of the progress on water resources management emerges from conferences and gatherings, from Lagos to Newcastle to the Istanbul Consensus. By bringing together different stakeholders and developing multilevel, multi-scalar networks, efforts to promote adaptive socio-technical responses to climate hazards can gather greater inspiration, motivation, and momentum. The time to act was yesterday. The most vital interests of humanity are access to clean water and protection from water-related disasters. Threats to those things bring us all together to create a better, more just, and equitable world.

# Index




H. Chang, A. R. Ross, *Climate Change, Urbanization, and Water Resources*, https://doi.org/10.1007/978-3-031-49631-8

**SPRINGER NATURE**

# GPSR Compliance

*The European Union's (EU) General Product Safety Regulation (GPSR) is a set of rules that requires consumer products to be safe and our obligations to ensure this.*

*If you have any concerns about our products, you can contact us on ProductSafety@springernature.com*

In case Publisher is established outside the EU, the EU authorized representative is:

Springer Nature Customer Service Center GmbH
Europaplatz 3
69115 Heidelberg, Germany

The manufacturer's authorised representative in the EU is Springer Nature Customer Service Centre GmbH, Europaplatz 3, 69115 Heidelberg, Germany. If you have any concerns regarding our products, please contact ProductSafety@springernature.com

Printed and bound by CPI Group (UK) Ltd, Croydon, CR0 4YY
01/10/2026
02225877-0016